Measurement in School Mathematics

1976 Yearbook

Doyal Nelson
1976 Yearbook Editor
University of Alberta

Robert E. Reys
General Yearbook Editor
University of Missouri

National Council of
Teachers of Mathematics

Copyright © 1976 by
THE NATIONAL COUNCIL OF TEACHERS OF MATHEMATICS, INC.
1906 Association Drive, Reston, Virginia 22091

Library of Congress Cataloging in Publication Data:

Main entry under title:

Measurement in school mathematics.

(Yearbook—National Council of Teachers of
Mathematics; 1976)
    Includes index.
    PARTIAL CONTENTS: Bitter, G. G. and Geer, C. A
metric bibliography.—Nonthematic essay: Hutchings,
B. Low stress algorithms.
    1. Mensuration—Study and teaching. 2. Metric
system—Bibliography. 3. Arithmetic—Study and
teaching. I. Nelson, Doyal. II. National Council
of Teachers of Mathematics. III. Series: National
Council of Teachers of Mathematics. Yearbook; 1976.

QA1.N3   1976   [QA465]   510'.7s   [513]   75–43533

Printed in the United States of America

Table of Contents

iii

Part Three: Selected Measurement Resources for Teaching

Part Four: Nonthematic Essay

Foreword

The year 1976 commemorates the 200th anniversary of the founding of the United States of America. This historic occasion provides an excellent opportunity to reflect over the years and think about how things used to be as well as how they might have been. Such reflection challenges us to think also about how things should be and to develop means of improving the current state of affairs.

The 1976 Yearbook, *Measurement in School Mathematics,* establishes a new birthdate for the National Council of Teachers of Mathematics as well—it is the firstborn of a new breed of NCTM yearbooks. As such, it is the first to reflect the new concept for NCTM yearbooks that was recommended by the Publications Committee and approved by the Board of Directors in April 1973. Both the character of the yearbooks and the procedure for producing them have undergone change. In particular, these yearbooks include four distinguishing features:

1. *The annual release* of an NCTM yearbook. Thirty-eight yearbooks (including the present one) have been published by the NCTM since its founding in 1920; yet, these have not been issued on a yearly schedule. Yearbooks will now be developed and produced so that a new yearbook will become available at each annual meeting of the Council.

2. *A central theme of current interest* to NCTM members. The 1976 Yearbook reflects a measurement theme. It provides ideas related to the mathematical and psychological foundations of measurement together with a wide variety of activities for learning and teaching fundamental notions of measurement. Subsequent yearbooks will focus on other themes holding keen interest among Council membership.

3. *A nonthematic essay* whose length, quality, and timely ideas warrant broad dissemination. The first nonthematic essay is entitled "Low-Stress Algorithms." It provides some insightful alternatives to traditional computational algorithms for both elementary and secondary teachers. Nonthematic essays will be a regular feature of NCTM yearbooks and will provide a forum for timely, high-quality manuscripts having broad appeal to classroom teachers.

4. *A grass roots approach* to the development of these yearbooks. No essays are solicited or commissioned in advance by the editors. A set of guidelines for the preparation of essays is developed, and resulting essays are submitted for competitive review. Each of the essays in this yearbook represents the results of several screenings and reviewings. This open competition for essays is unprecedented in the development of NCTM yearbooks, providing the opportunity for fresh ideas and different perspectives that might otherwise go untapped.

Although a new yearbook concept is being generated, NCTM members will be pleased to know that the more definitive reference works (previously known as yearbooks) will continue to be produced. This professional reference series will maintain the tradition of high quality as well as the comprehensive treatment of specific topics that has been established in previous yearbooks.

The production schedule means that each yearbook, including this one, must progress from conception to completion within approximately two years. This is an ambitious schedule that demands cooperation. Essay writers, reviewers, editors, and the NCTM production staff must meet their deadlines if the yearbook is to be produced on time. *Measurement in School Mathematics* is evidence that such cooperation exists and can be mustered.

The 1976 Yearbook would have been impossible to produce without substantial help from many sources. In addition to producing essays of high quality, the contributors met tight deadline schedules and were most cooperative in all stages of development. Thanks are due to each of them for their work as well as their patience.

In addition to writers, earlier yearbooks have had editorial panels to monitor the various stages of production. The 1976 Yearbook had no official editorial panel; yet many individuals have served in various editorial capacities, and their labors should not go unnoticed. Members of the Publications Committee—including Arthur F. Coxford, Donald H. Firl, Larry L. Hatfield, Shirley A. Hill, Thomas E. Kieren, Jeremy Kilpatrick, and Seaton Smith—as well as NCTM presidents Eugene P. Smith and

E. Glenadine Gibb have made many significant contributions and were instrumental in molding this book. With no editorial panel to call on, the screening and reviewing of proposed essays relied heavily on feedback from individuals, many of whom were mentioned above. In addition, Douglas B. Aichele, Jack Bana, Raymond A. Cornett, W. George Cathcart, Edward J. Davis, John B. Dubriel, Douglas A. Grouws, Joan E. Kirkpatrick, Thomas R. Post, and Paul Wm. Rahmoeller provided valuable new insights and many helpful reactions, comments, and suggestions. Sincere appreciation is also due to Charles R. Hucka and the production staff of the NCTM for their help in making this yearbook a reality.

As first general editor of NCTM yearbooks, it has been my pleasure to work with Doyal Nelson, issue editor for *Measurement in School Mathematics*. Doyal did the lion's share of the work, and he did it extremely well. He was instrumental in all stages of development and consistently, yet discreetly, kept us on the right track. Together with the help and assistance of many, we have moved toward the final stages of this yearbook. It has been a challenge that we have answered to the best of our abilities. It is our hope that this yearbook will justify the faith that both the Publications Committee and the Board of Directors have in the new approach to yearbooks.

ROBERT E. REYS
General Editor

Preface

The introduction of the international system of measures (SI) as the basic measurement system in the United States and Canada was a major factor in choosing measurement as the theme for the first yearbook in this series. In anticipation of the introduction of SI, mathematics teachers and others have been active during the past few years in creating materials designed to facilitate the introduction of the system. It was hoped that this activity would be reflected in the yearbook. The aim, however, was neither to produce a treatment of how best to introduce new systems of measure in the schools nor to produce a definitive treatment of measurement in general. Rather, our intention was to generate a series of essays on current issues and interests covering the whole spectrum of measurement and its role in school mathematics. Current interest and timeliness were the major attributes sought after in calling for and selecting the material that appears in this yearbook.

When the call went out for contributions on the measurement theme, it was expected that the specific areas chosen by the various contributors for discussion would be wide ranging, for there was no preconceived notion of how the book should be organized. Once the material had been chosen, however, three broad areas into which it might, for reader convenience, be divided became apparent. The measurement material of the book has, therefore, been partitioned into three sections, as indicated in the Table of Contents.

The essays in Part One address themselves to a number of divergent issues and problems associated with learning and teaching measurement ideas. Sanders uses the continuing achievements in space to illustrate the vital role measurement plays in man's activities. Osborne reveals how transfer of learning can be facilitated through a careful consideration of the mathematical foundations of measurement. The vagaries of children's thought in measurement situations is treated by Steffe and Hirstein as they review current research into children's conceptual development. Inskeep outlines a plan for developing a program of activities designed to enhance children's understanding of measurement and their ability to deal with measurement situations. The importance of estimation as an aspect of measurement and how it can be treated in school mathematics is the theme of Bright's essay, whereas Kerr and Lester propose a specific routine for helping children to understand and deal with measurement error. Finally, some vital problems and issues to be faced while SI is being introduced as a basic system are identified and discussed by Sawada and Sigurdson.

Measurement can probably provide some of the most novel and attractive activities in school mathematics. Part Two contains a selection of some unusual ones. A clever device that Kullman calls the Model T range finder can be used effectively by fourth graders and at the same time provide high school seniors with some challenging geometry. Rahmoeller has outlined a method of using photography for measuring the speed of rapidly moving bodies, and Smith and Rachlin reveal the intricacies of using the carpenter's square. Beem lays the groundwork for a series of pupil experiments with the measurement of surfaces. Out-of-door measurement activities constitute Gordon's central theme; here again are suggestions for activities that will appeal to a wide range of abilities and ages. Schupp outlines the construction of a device for measuring the speed of a projectile and leads the reader into an interesting analysis of measurement error.

The material in Part Three is not in essay form. It is intended to serve as a bibliographic source of materials and ideas for the teaching of measurement. The first part, prepared by Jackson and Prigge, includes photographs of useful devices for teaching measurement ideas as well as suggestions for their use in actual classroom situations. In the second part, Bitter and Geer have provided an annotated list of materials focusing on a metric theme.

Our instructions were that this yearbook should also contain some nonthematic material that we believed would be of immediate interest to our readers. Part Four, then, is a nonthematic essay, which we hope will appeal to all mathematics teachers. Hutchings outlines a method of teaching computation employing what he calls low-stress algorithms. These low-stress algorithms provide an alternative to traditional computational algorithms. As such, they are designed to relieve the child of the memory burden imposed by most algorithms in use today.

It is not necessary to start with the first essay and then read them all in order. Although it may be useful to read a group of essays together, our intent was to make each essay as completely independent of any other as possible. We wanted the reader to be able to look at the Table of Contents and begin at any one of the essays according to interest.

Our wish is that this first yearbook of the new series will be useful and informative. We also hope that when a call goes out for contributions to future yearbooks, many of our readers will be stimulated to respond.

DOYAL NELSON
1976 Yearbook Editor

1

Why Measure?

Walter J. Sanders

On 20 July 1969, Neil A. Armstrong stepped out onto the surface of the moon. Do you remember what you thought or said at that moment in history? Perhaps you reflected on the bravery of those men who made that "giant step for mankind," or marveled at the scientific technology required to perform such a feat, or swelled with pride in your country, or maybe you prayed that they would get back to Earth safely. But chances are you did not think about measurement. It is hard to imagine anyone at that moment thinking about it as a marvelous feat of measurement. Yet, this incredible achievement would not have been possible without measurement.

Measurement behind the Scene . . .

In spectacular achievements

Hidden in most of humanity's spectacular accomplishments are innumerable measurements, each related to or dependent on myriad other measurements. Thus, for a moon shot the navigational problems alone require many highly precise measurements of distance, time, duration, relative position, size, weight, and speed. The spacecraft must achieve and then

Space Flight. Astronaut Edwin E. Aldrin, Jr., walking on the surface of the moon, 20 July 1969. (Photo courtesy of the National Aeronautics and Space Administration)

escape the earth's orbit, achieve lunar orbit, land the lunar-earth module (LEM) on the moon, retrieve the LEM, and return to a splashdown in a specified region on Earth. For such maneuvers the correct time of ignition and duration of burn are necessary. These measurements in turn depend on the weight and size of the space vehicle.

Preparations for a lunar landing also require many measurements of size, shape, and weight in the construction of the space vehicle and the related ground-support complex. Although the millions of parts are manufactured by many companies in various parts of the country, each part must fit with the others. A booster stage built in California, for example, must fit with the other components of the space vehicle when it is assembled at the John F. Kennedy Space Center in Florida. This can be assured because the components are made to carefully measured specifications. Each launch vehicle is capable of placing a payload of specified size and weight into earth orbit, and the spacecraft is designed so that it does not exceed these limitations. The silo, gantry, transporter, and launching pad are all designed to accommodate the physical characteristics of the space vehicle. The many interrelated components function together because of the magic of measurement.

The spacecraft can be thought of as a very large suitcase in which everything necessary for the complete space flight must be packed. Size and weight limitations demand that careful consideration be given to each item of equipment that is included. Necessary fuel, food, water, air, and communications and navigational gear are a must—as are the crew, of course. Scientific instruments for gathering data also receive a high priority. All other equipment must be carefully considered in terms of its size, shape, weight, and value.

In everyday life

What does all this have to do with John and Jane Doe, private citizens? They will be participating more fully in the space age if they understand the role of measurement in space exploration, and, ultimately, they will enjoy new products or services made available for general use by the advances made in technology. But more important, the underlying technology that made the moon shot possible is the same technology that governs so much of our everyday life. If people had not developed the measurement processes, on which modern technology is based, our way of life would be much like that of lower animals. One would not even be able to go to the supermarket and buy a can of tomatoes. In fact, there would not be any supermarkets to patronize. Let's see why this is so.

The companies that process the canned goods sold at a supermarket would not be able to guarantee that their product would not spoil when sealed in the cans without measuring the temperature at which the food is cooked and without measuring how long it is cooked. The cans themselves are produced by machines made up of interrelated parts that must fit together within certain tolerances; otherwise, the machines will not operate. Moreover, the metal sheets from which the cans are made must be of the proper thickness if the machine is to work successfully.

A supermarket retailer operates as a middleman; that is, goods are purchased from various suppliers and made available to the consumer. What type of an agreement is made between the canning company and the retailer? The retailer agrees to buy the food at a certain price per can. This cost is added to the expected overhead and profit, from which a retail price is established. The transactions are made with money. But wait a minute: pricing and the use of money as an exchange medium are based on the measurement of value, and so in a world without measurement, money would not be available. Thus the retailer and the canner must negotiate a deal by bartering. Each participant must bring real goods to the negotiations, and when both parties agree that the piles of goods are of equal value, the exchange is made. How much more convenient it is to use money.

After the marked items have been selected at a supermarket, they are taken to the check-out counter and the clerk is paid the total of the prices

Bartering. "I'll trade you my bead frame for your ball."

stamped on the items, plus the tax. (How often was measurement used here?) Imagine how complicated the check-out procedure would be if each customer had to barter rather than pay with money.

A supermarket operates by distributing goods on an equitable basis to its customers. It does this by measuring out equal portions and offering them for the same price. This permits a larger volume of business, which results in lower prices. Without measurement, then, there would be no supermarkets.

Suppose you get into your car to take a trip and notice that you are low on gasoline. You stop at a service station. How do you settle with the attendant for the gas you receive? The gas is measured as it is pumped into your tank, and the cost is measured simultaneously. Since two service station customers rarely require the same amount of gasoline, it would be very difficult to deal with customers fairly without using measurement.

Automatic Gas Pump. The measurements are made automatically for you.

In a world without measurement, however, you probably would not have an automobile to drive and would not have to worry about getting fuel for it. Measurement allows us to produce cars and other manufactured items, such as television sets and washing machines, in mass, which makes them available at accessible prices. In mass production, each part of a product is made in large quantities and carefully measured to size so that any copy of a given part will fit in the final assembly of the product. Furthermore, if a part becomes defective after the product has been sold, a replacement part can be ordered, and the part will fit. None of this would be possible without measurement.

Measurement and the Individual

Because people do know how to use measurement, we have relatively inexpensive manufactured goods, a system of monetary trade, space travel, and all those other benefits of measurement. But do I as an individual need to bother learning how to measure? Can't I let other people worry about making the necessary measurements? After all, when I shop, I just take items off the shelves, put them in my shopping cart, go through the check-out line, count out (not measure) the correct change (or let the clerk do this for me), and take my purchase home. If I need fuel for my car, the pump automatically measures the volume of gasoline and the cost, and I pay for it with my credit card. At the end of the month I receive a statement for the total charges and write a check for the amount of the billing. I can accomplish all these tasks even though totally ignorant of the measurement process.

It is possible for a person to go through life without making measurements or being aware of the concept of measurement. Animals are surely ignorant of the intricacies of measurement, but they manage to live in the same world, reaping some benefits of measurement. For example, a pet dog is probably fed a specially prepared commercial dog food in which the ingredients have been carefully measured to assure a proper balance of vitamins and minerals for optimal nutritional value. The dog profits from measurement even without an appreciation of it.

Measurements Everyone Makes

Is it to our advantage to avoid measurement? The following examples suggest how the ability to measure helps us enjoy a full, richer life.

Telling time

We measure time in order to know when to leave for school, when to meet a friend for a date or appointment, when to arrive at the theater for the start of the feature, when to turn on the television for our favorite program, or how long to boil our egg or heat our TV dinner. It would be very inconvenient to rely on someone else always to do these things for us.

Placing an order

If we wish to fence in our back yard, we might call a local carpenter and leave the entire project in his hands. Or, to save money we might order the fencing and do the job ourselves. Here we would need to be able to tell the merchant how much fencing we need. This can be accomplished by measuring the length and width of our yard and determining by a simple calculation the amount of fencing needed. Often we need not even leave our home because we can order the fencing by phone or mail and have the store deliver.

New draperies for our picture window may be ordered through catalog sales. To order them we need to give the choice of material, the length of the window, and the height above the floor for the top of the drapery. If we are replacing the draperies on several windows and are going to make them ourselves, we shall need to make more measurements, including those which help us cut the material in the right places.

When we decide to paint our own house, we need to know how much paint to buy. The label on a can of paint usually tells how many square feet of surface the paint in the can will cover with one coat. So in ordering paint, we should first measure the dimensions of all the rooms to be painted, calculate the area, and order the required amount of paint for the number of coats we plan to apply. Many homeowners have not bothered to make these measurements before purchasing their paint and have run out of paint before the job was finished, only to find out that their particular paint has been discontinued with no more available to finish the job.

Mail orders placed through a catalog often depend on measurements. The size of a piece of furniture may be what determines whether it is ordered, since it will need to fit in a particular space in the living room. A tablecloth is ordered to fit a particular table. Sheets are ordered to fit a bed of a particular size. Most department store catalogs devote about half their space to clothing, all of which must be ordered by size.

Making comparisons

Two mountain peaks in the state of Washington are Mount Saint Helens and Mount Rainier. Which of these two peaks is the taller? How could you find out? You certainly cannot move them side by side and compare their heights as one would two people, but their heights above sea level can be measured so that a comparison can be made.

In this example, two objects whose heights were to be compared were remote from each other, making direct comparison difficult or perhaps impossible to carry out. Measurement offers the only reasonable means to complete the task. There are many examples like this. Determinations of

Mount Saint Helens with Mount Rainier in the background. It is difficult to determine which peak is the taller from this view. (Photo courtesy of the United States Department of Agriculture, Forest Service)

the world's largest lake, the tallest waterfall, or the deepest spot in the ocean are typical. These problems are solved by specialists. Other examples are important to ordinary people. Suppose you are taking down your storm windows, but your ladder is not long enough to reach an attic window. The only neighbors having a long ladder live three blocks away, but you are not sure their ladder is any longer than yours. Do you walk to their house and carry their ladder back to find out if it is longer than yours? Or do you measure yours first, then measure theirs (or ask them to do so) to find out if it will work before you carry it all the way back to your house?

Sometimes the comparison in which you are interested is between events that occur at different times rather than in different places. For example, in a race there may be too many entrants to have them all run at the same time; so they are separated into two heats (half the entrants run the race, and when they finish, the remaining contestants run). But if two contestants are in different heats, how can they be compared? Usually each is timed and the results compared. Most records in athletic events are determined in this manner. A record for a race (running, swimming, motoring) is given in terms of the distance and the time. Other records are given in terms of the distance only, as in the discus throw or the high jump. A fishing record may be given in terms of the weight or the length for a given type of fish.

The weather information recorded daily allows one to compare the weather on a day-to-day basis or make a comparison between years. If a farmer tests a new hybrid in a year with unusual weather, the results may be misleading. The crop may have been great because the weather was ideal for it, whereas the next year more typical weather will produce a poor crop. A rising or falling barometer may help a mother decide how to dress her children for school.

Teachers often compare their classes with previous ones. Memory is usually imprecise in this matter. Standard measures of achievement are usually much more reliable. Teachers also like to know how their students compare with the students in other states. To find out, they measure their students' achievement and compare it with a similar measurement of the other students'.

Mixing ingredients

Not everyone mixes up batter for a cake or mixes tints in several cans of paint to produce the same color, but most people find occasion to measure ingredients carefully for some type of mixture. If you drink coffee, tea, or frozen orange juice or limeade, a more satisfying drink results from a careful measurement of the product and the water than from a haphazard mixture of the two. If you mix your own starch and make the mixture too thick, the clothes will be too stiff. The wrong mixture in preparing concrete may result in a product that doesn't set properly or that crumbles easily.

Dosages

Each drug on the market has been carefully researched on its effectiveness in specified quantities for people of designated ages. Special care is taken to establish safe limits to prevent an overdose. When the drug has been prepared in pill form (in which case the drug company has precisely measured a specified dosage), the patient need only count the correct number of tablets to take; but often the medication is in liquid (or powder) form, and so the patient must measure out the correct dosage. In a matter such as this it behooves the patient to be skilled in measurement.

Equitable sharing

It is easy to break a candy bar into three pieces of nearly the same size to share among three people, a task that can usually be done without careful measuring. Separating a large cattle ranch into three equal-sized parts without measuring, however, would be extremely difficult.

Suppose two families are to share a large box of apples. If the apples are of uniform size, equitable shares can be determined by counting. But if they are of odd sizes, one family may get most of the smaller apples.

A fairer distribution can be achieved by weighing. One family may get more individual apples than the other, but each could make the same amount of apple sauce. This is why supermarkets often sell produce by the pound rather than by the count.

Avoiding counting

Large quantities of nails, bolts, washers, or nuts are bought by weight or by volume rather than by counting out the number of items, since it is so difficult and time-consuming to count them but easy to measure them. If nails are sold by the box, the average number per box is easy to determine and can be printed on the box. The same is true for weighing. Measuring gives a quick, fair way to sell these items.

Determining amount by weighing is useful in situations other than those that would ordinarily be done by counting. For example, wire is usually bought by the kilogram or some other weight unit rather than by length. Can you imagine how difficult it would be to measure the length of wire on a spool? It is much easier to determine the weight per meter and the total weight, and then calculate the approximate length.

Determining temperature

When you call a doctor's office because you are not feeling well, the first thing the nurse asks you is, "Have you checked your temperature?" Body temperature is an excellent indicator of infection, and many times knowledge of one's body temperature may be the deciding factor in whether or not to contact a physician.

Evidence indicates that room temperature is important for physical well-being. Teachers need to provide a healthful temperature in their classrooms, and every family needs to control the temperature of their home for optimal health. Furthermore, excessive heating of rooms wastes fuel. It is in the national interest for each citizen to maintain a home temperature of 20 °C during cold weather.

Familiarity with the daily temperature can help parents decide how warmly to dress their children for school or help car owners decide whether they need to add antifreeze to their car's cooling system.

Summary

Many things that we do are possible because of measurement. We may not need to do the measuring ourselves, but someone else does. Without measurement there would be no cheap manufactured products, no scheduled public transportation, no aircraft flights at night or in bad weather, no use of money for exchange purposes, no space travel, and a greatly reduced body of scientific knowledge. The world as we know it is possible because of measurement. The answer, then, to "Why measure?" is simple: Measurement is necessary if we are to have the things we wish to have and to do the things we wish to do; the luxuries to which we are accustomed and do not wish to do without are a product of modern technology, which is totally dependent on measurement; and many of the simple tasks we all perform and take for granted are made possible or easier through measurement.

2

Mathematical Distinctions in the Teaching of Measure

Alan R. Osborne

The child has experiences in early elementary school with measures of time, length, area, and liquid. Progressing through school, the student encounters such diverse but representative measure systems as weight, direction, volume, pollution indexes, test scores, energy, and light intensity. Each of these measure systems possesses one or more characteristics differentiating it from other systems of measure. But these systems also share one or more common attributes. Learners do encounter and use these common and differentiating attributes within a variety of types of measure systems both in and out of the classroom.

Given the diversity and extent of students' experiences with measure, the critical question is, Why do students have so much difficulty with measure? One would expect that students would find it progressively easier to learn about measure systems and would apply their understandings more readily as their experiences accumulate. However, this does not appear to be the case. Beginning physics courses in college typically include a major unit on measure. Witness the discomfort of many individuals when they contemplate conversion to the metric system—a stark testimony of the failure of many people to attain success in coping with measure. In short, the critical question concerning the curriculum and teaching of measure is, Why do students show so little evidence of using their prior experience with measure?

The question is one of transfer of learning. Given the number of different measure systems a student encounters and the extensive use of measure concepts and ideas in mathematics, science, and most vocations, it is easy to see that a teacher's primary goal in designing measure experiences is to facilitate transfer. The learning within one measure context should aid the learner in the subsequent learning and use of measure concepts, principles, and generalizations.

The purpose of this essay is to analyze the teaching of measure from the standpoint of designing instruction to promote transfer. In the next section some powerful ideas concerning transfer are discussed. Subsequent sections examine the mathematical bases of measure in terms of designing instruction to facilitate transfer.

Transfer

We do know a lot about teaching for transfer. The research of educators and psychologists, as well as the experiences of countless classroom teachers, provides sound guidance for the design of instruction to facilitate transfer.

Transfer assumes that two sets of concepts, principles, and generalizations are to be learned. These sets must share some common characteristics and attributes. The learning of the first set P of concepts, principles, and generalizations (prior learning) must precede the learning of the second set S (subsequent learning). Positive transfer occurs if the learning of P enhances, improves, or makes more efficient the learning of S.

Six principles for designing instruction to facilitate transfer are stated below. They are based on the assumption that transfer is a product of the learning of P. That is to say, transfer depends on both *what* is learned and *how* it is learned. If the instructional approach for volume is based on the prior learning of area concepts and generalizations, then the potential for transfer hinges not only on what the student knows about area but also on the stimuli and mediating processes used to establish the area concepts and generalizations. Thus, the teacher must design instruction for both P (area) and S (volume) to facilitate transfer. Transfer is not achieved simply by stating to students learning about volume, "Do you remember when we were studying area, we learned . . ." Instead, teaching for transfer demands the teacher be Janus-like: the teacher must look ahead to ideas and concepts yet to come, as well as look back to those concepts established prior to the day's instruction.

The six principles for designing instruction to achieve transfer are generalizations based on psychological evidence of how people learn. In one sense they are oversimplifications of extremely complex learning processes; nevertheless, they are to be considered maxims that will be useful in planning for teaching.

Principle 1. The teacher must identify those attributes of learning common to P and S. These attributes include the products of learning (concepts, principles, and generalizations) as well as the conditions for learning.

The systems of measure for both length and area share the property of additivity. That is, for two segments of a line, $\overline{AB}$ and $\overline{BC}$, length $AB +$ length $BC =$ length AC. In the same way, the area of a polygonal region can be found by adding the areas of subregions I and II of the polygon as indicated in figure 2.1. The principle of additivity is, therefore, an attribute

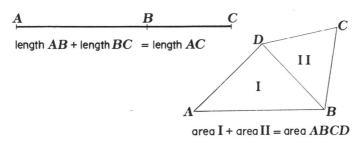

Fig. 2.1

of learning common to both P (length) and S (area). It is an outcome, or product, of the learning. An identification of the conditions of learning would include those common types of problems or stimuli from which the child builds the concept of additivity. For example, a teacher might be concerned that the child not apply additivity to overlapping segments, say, length $AB +$ length AC where B is between A and C. The teacher should design conditions of learning or problem situations where this does not work so that the child's inclination to add them will be torpedoed. Comparable, parallel activities need to be identified for area.

Principle 2. The instructional materials and design for learning P must emphasize or make explicit the attributes of P that are common to P and to S.

The previous example for length (P) and area (S) suffices. Once the attributes are identified, instruction for (P) must establish the attribute.

For additivity, instruction for length (and distance) concepts must be designed to assure that students acquire the principle. Otherwise, there is no base for the transfer when the student encounters the area system of measure.

Principle 3. The more complete and thorough the leanring of P, the greater the likelihood of transfer.

In many senses, this principle is a reaffirmation of principle 2, except that it is stated in terms of student learning rather than the design of instructional materials and experiences. In other words, if the teacher's classroom evaluation and testing reveal that students do not understand additivity for length, then the teacher cannot presume to use it as a base for designing transfer activities for the additivity property within the context of area.

Principle 4. The design of instruction for S must be in terms of the attributes and conditions for learning that characterize P.

Principle 4 states that if the base for transfer has been established for the prior learning (*P*) and if the teacher has designed instruction for *S* in terms of this base of a common attribute, then the learning of *S* is more likely to occur. In the length-area example, if the additivity property for area is the goal of transfer of learning, then the design of learning activities for area should be directed to that goal. If instruction for *S* is not pointed directly toward the common attribute, then the probability of transfer is decreased.

Principle 5. More powerful and inclusive concepts, principles, and generalizations have greater potential for facilitating transfer than less powerful and less inclusive ones.

The idea embodied in principle 5 is familiar ground for mathematics educators. Restated in familiar terms, it maintains that unifyng concepts, such as those delineated in the Twenty-fourth Yearbook, *The Growth of Mathematical Ideas K–12* (NCTM 1959), serve best as a basis for designing instruction for the goal of transfer. Thus, instruction for transfer would be best when it is based on attributes familiar to the mathematician as powerful ideas, such as function.

Principle 6. Instruction for S must highlight the differences as well as the similarities of the attributes of P and S in order to protect the learner from overgeneralization and misapplication of the concepts, principles, and generalizations of P to S.

The highlighting of differences is an essential component in planning instruction for transfer. It is a warning to be wary of those situations in which a student is inclined to transfer learning where it is inappropriate. For example, in the length-learning (*P*) and area-learning (*S*) situations, students working with length soon learn that if two segments are congruent, then they have the same measure. They also learn that the converse is true, namely, that equal measures for segments implies the congruence of the segments. With area, however, only the attribute of congruence implying equal measure holds. One cannot argue that because two rectangles have area 8, they are congruent. Thus, in designing instruction for transfer, the teacher must protect the learner from shifting too many of the properties of the prior learning P to the new learning situation S.

These six principles may be used as prescriptions for designing curricula and instruction concerning measure. They indicate that transfer is a process requiring careful planning. An essential component of this planning is an analysis of the content in order to identify the "stuff" of transfer, namely, the attributes common to both the prior and the subsequent learning. The analysis should also identify those attributes and characteristics that differentiate the two sets.

The following sections will include a careful examination of the mathematics of measure to help establish the basis for designing instruction for the goal of transfer. First, the nature of measure is discussed, and then several different systems of measure are examined and compared.

The Nature of Measure

Students can become confused about the nature of measure. Some of their instruction deals with measure in science and some with measure in mathematics. In teaching mathematics, the teacher appropriately stresses that corresponding to two given points there is a single, unique number that is called the distance between the points. In science class, though, the student has studied relative error, precision, and accuracy, perhaps while determining the length of a spring. In the contexts of both the mathematics and the science, the common words *length, distance,* and *measure* have been used. But two tasks that ostensibly are the same, finding lengths, have quite different outcomes. In one case the outcome is a single number; in the other, it is a range of numbers. Thus the student faces an apparent ambiguity—exactness in the mathematics of measure but inexactness and imprecision for measure in science. Exacerbated by the common words and parallel processes being used in both settings, the ambiguity often leads to unstated confusion.

The ambiguity and confusion need resolution at some point in a student's career if only because the recognition of the reasons for the ambiguity reveals so much concerning the nature of science, of mathematics, and of measure. For the scientist, measurement is a cornerstone of the observational processes of doing science. Observation and measurement are used for two purposes: first, to build a model of reality, and second, to test the truth of the models of reality. Thus, measuring is both a process and a skill. Helping children use and understand a centimeter scale or a protractor is helping them acquire an observational skill and a process. A measurement, then, is the result of observing. It can be incorporated into the model or used to test the truth of statements in the model.

The difficulties inherent in the observational act of measuring lead naturally into the concepts of precision, accuracy, and relative error. The learner faced with the decision of determining whether the segment in figure 2.2 is closer to 8 cm or 9 cm can appreciate the necessity for saying

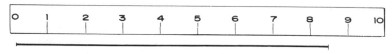

Fig. 2.2

there is error in the measurement process. Activities and conversation about the measurement act should highlight the fact that error can develop in many ways. For example, the importance of accurately determining or placing the zero point should be stressed as a skill applicable in many measurement settings. The difficulties some children have in aligning a protractor properly may be decreased if the appropriate transfer base has been established in the length setting. Facing children with the conflict of deciding whether a measure is closer to one of two scale coordinates leads naturally to an appreciation or receptivity for precision. (See figure 2.3.) The learner's resolution of the conflict should have two outcomes:

1. Measurement is an observational decision by the measurer to classify a segment to the nearest unit.
2. The way to improve precision is to improve the measuring device or the process.

The first outcome is the equivalent of the standard textbook definition of precision of measurement. The child's typical observation that he can "see better" than the ruler is scaled establishes the possibility and the need to subdivide the units on the scale further in order to improve the precision of the measure.

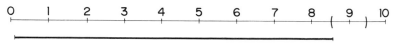

Fig. 2.3

The learner's observation that measurement processes necessarily involve judgment and error needs to be accepted by the teacher and capitalized on to help the learner build an understanding of the differences between measures in a scientific setting and within the context of the exactitude of mathematics. Some of the judgments are observational decisions, but others are more a matter of values. For instance, relative error is typically explored at the junior high school level by defining it as the ratio of the greatest possible error (half the unit) to the measurement under consideration. This definitional approach is correct but suffers by being somewhat removed from the reason the scientist is interested in relative error. The distance from the Saint Louis Arch to the Washington Monument could be measured either to the nearest centimeter or to the nearest kilometer. There would be a difference in the relative errors. The value judgment concerning the relative errors is largely a matter of the purpose in making the measure observation. For the purposes of the tourist traveling from the Washington Monument to the Saint Louis Arch, the measurement of the distance to the nearest centimeter is both foolish and expensive. The definition of relative error is a quantification useful in making the value judgments of the relative merits of measuring at different levels of precision or accuracy. Too frequently the learner copes with the definition on a formal, definitional basis without considering the behind-the-scenes value judgments of how accurate the measurements need to be to fit the task at hand.

One of the more important ideas a learner can acquire about error is its effect on computation. If a measure is multiplied by a constant, the error in that measure is also multiplied by the constant. If two measures are multiplied, the product must be used with consideration of the effect of the error. Thus, for the computation of the area of a rectangle with sides 58 cm and 20 cm, we have $(58 \pm .5) (20 \pm .5)$ cm^2, or an area measure between 1121.25 cm^2 and 1199.25 cm^2. Any addition of measures needs to be examined in terms of effects of error also.

Learners are quick to note that most instruction in mathematics concerning measure systems does not require that error be taken into account in computation. Most texts, and many teachers, do not bother with error factors in computation except in those special sections of the text concerning accuracy and precision. Consequently, many students dismiss error as unimportant. In the practicalities of measure, error is important to the scientist and engineer. In mathematics, error helps the learner distinguish between the use of measure in science and its use in mathematics. That is to say, error and observation need to be contrasted with the exactness of measure in mathematics.

The mathematician's primary concern is operating within a model or within an ideal world independent of reality. Measure is an entity rather than a process. The mathematician's test of truth is the correctness of reasoning, not whether the model fits reality. The consistency of the model or the ideal world is of supreme importance. The mathematician operates within the model by providing the syntactical rules, or "grammar," governing the reasoning within the model. Measures are the nouns within the statements formed or operated on within the grammar rather than the verb form, which would be indicated by the observational process of measuring. Although the scientist behaves like a mathematician while working within the model and is sometimes concerned with how well the model fits reality, measure is different for the two.

In the remainder of this essay, the word *measure* will be used to indicate the mathematical use of measure independent of the observational factors of imprecision and error. The words *measurement* and *measuring* will indicate observation and scientific use. Figure 2.4 illustrates the distinction.

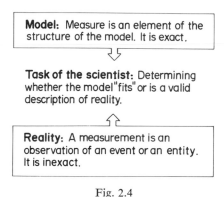

Fig. 2.4

The idea of model building is one of the most powerful ideas of modern sciences. There can be many descriptions of reality. For example, the Newtonian model of the universe is practical and useful for some scientific purposes, whereas the relativity model is desired for others. Through the careful examination of measure processes from the standpoints of both science and mathematics, some of the characteristics of models and modeling can be considered.

The idea of a model is relatively sophisticated. The teacher must recognize whether measure is being used in the scientific or in the mathematical sense. Children at the early learning stages cannot appreciate the distinctions. Indeed, we know that most of the successful means of establishing basic number concepts depend on manipulative activity with physical objects. We establish with the young child the expectation that physical reality is a part of mathematics, a view inconsistent with the model concept previously described. A careful treatment and analysis of measuring as an observational process is critical for more mature learners if they are to understand modern science.

Children need to consider measure from the standpoint of both the observational system and the model. The former is critical not only to understanding science but also is the type of understanding most practical in adult life as the proverbial "man on the street." Children use the observational processes to quantify their world and to make mathematical and verbal descriptions of their environment. The mathematical understanding of measure within the intact system or model is necessary for building future mathematical concepts. Each understanding plays a significant role in establishing a base for transfer. Establishing the distinction between the mathematical and the scientific use of measure decreases the probability of confusion arising from negative transfer.

Functions and Measure

Every system of measure is based on a function. The most powerful idea characterizing measure is the concept of measure as a function. It should be at the heart of designing instruction for transfer. It is the unifying idea that brings all measure concepts and principles into focus, whether considered from the standpoint of science or mathematics.

The properties of the measure function characterize the measure system. A function associates a domain element with a unique range element. For the scientist determining the effect of temperature on the expansion of a steel bar, the various temperature states of the bar constitute the domain set. Corresponding to each temperature state is a range element, an interval of numbers that is the result of an observation. The idea of function is so thoroughly intertwined with measurement in a scientific setting that the function idea provides the primary basis for transfer.

We now shift to consider measure in a purely mathematical sense. Several measure systems: length, area, volume, and angle measure, will be explored.

Length

Length is one of the first systems of measure the young child encounters in school. It is fitting, therefore, that the mathematical nature of length be examined for its potential as a base for the transfer of learning to other systems of measure. Transfer assumes there are two sets of learnings, one prior and the other subsequent. Thus, the reader should expect several ideas and concepts that are foundational to the measure of length to reappear in different forms as other, subsequent systems of measure are discussed.

Length in the world of mathematics is exact. Corresponding to two points A and B, there is precisely one number that is the distance from

A to *B*. That is, distance is a function that maps segments of a line into the set of real numbers. We define the distance function *d* in terms of a set *M* of segments on a line. The function maps the elements of *M* into the set of nonnegative real numbers. We write

1(*a*). $d: M \rightarrow \mathcal{R}^{+,0}$, or

1(*b*). $d(A,B) \in \mathcal{R}^{+,0}$, where *A* and *B* are the endpoints of the line segment.

This association of line segments with positive real numbers is the foundation of length. The notation is complex and quite different from that encountered by a child in elementary or junior high school. Distance is typically labeled *mAB* or *AB* by the junior high school years. The usual elementary treatment of length is in terms of exhibiting the numbers that are the lengths. This development uses the more precise notation because it allows the .identification of the foundational ideas that characterize the distance function.

The attributes or characteristics of the distance function that provide the bases for transfer can be seen in the relationships between the elements of the domain and the range of the function. Indeed, these are the primitive concepts the young child must acquire to understand distance. Some of the ideas are obvious, so obvious that it is easy for the adult to overlook the fact they need to be established with children. Each property is first stated in functional notation, then it is restated in less precise language.

2. $d(A,B) = d(B,A)$. This states simply that the distance from a first point *A* to a second point *B* is the same as the distance from the second to the first.

3. If $d(A,B) = 0$, then $A = B$, and conversely. The child can reason that if the distance between two points is zero, then the two points are identical. The "and conversely" means the child can reason in reverse, namely, that the distance from a point to itself is zero.

4. The additivity property: If *B* is between *A* and *C* on a line, then $d(A,B) + d(B,C) = d(A,C)$. This states that the length of the union of two segments laid end to end on the line is the number that is the sum of the lengths of each segment.

5. The congruence property: If segment *AB* is congruent to segment *CD,* then $d(A,B) = d(C,D)$, and conversely. That is, congruent segments are the same length, and if two segments are of the same length they are congruent.

6. The Archimedean property: If *B* is between *A* and *C* on a line, then a counting number *n* can be found so that $n(d(A,B)) > d(A,C)$. That is, if a segment *AB* is contained in segment *AC,* as indicated in figure 2.5, then a

positive whole number n can be found that when multiplied times the length of AB is at least as large as the number naming the length AC. It may be more intuitive to say that if copies of segment AB are laid end to end in the direction of C, an end of one of the copies will eventually fall on or beyond C. The number of copies of length AB is the number n. This iteration provides some of the basic understanding necessary for learners to develop a coordinate system for points on the line.

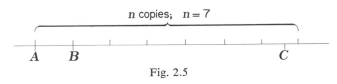

Fig. 2.5

Taken together, the six properties describe distance on a line. Usually, number is integrated into the distance function by adding a coordinate system to this set of ideas. This turns the line into a number line. Another function is used to map the real numbers onto the line in a one-to-one fashion. If the coordinatizing function is called K, then A is labeled with the number $K(A)$ and B with $K(B)$. Distance is then defined by $d(A,B) = |K(A) - K(B)|$, where the absolute value is used to assure a positive number for distance.

These six properties, coupled with the coordinatizing of the line, constitute a very abstract analysis of the measure of distance on a line. Fortunately, the experiences of children with distance are not so abstract. Each of these ideas needs to be established—and indeed overlearned—if the child is to have a good base for transfer. Children must, for example, encounter numerous situations in which they have a chance to use the additivity property. These situations may be as commonplace as measuring the distance across the classroom on a line and comparing the result to the sum of measures from each wall to a certain point on the line. Or it may be in terms of the number-line approach to building addition facts. The teacher has the task of designing activities directed to each of these primitive subconcepts for distance.

The child typically has the majority of experience with distance in a computational setting, usually one that stresses formulas. The ideas embodied in the properties are considered together rather than established separately. Is this the best way to establish these six properties of distance? The answer is not known with certainty. It can be said with certainty, however, that in terms of looking ahead for transfer, each of the six properties must be established as important.

A notable characteristic of the six distance properties is that they are stated in terms of segments on a single line. Although this is somewhat artificial, many of the preliminary experiences of children in the early years of elementary school deal with the number line or the use of a ruler in a one-dimensional setting. Children do need, however, to encounter length measure in the more natural two- and three-dimensional settings. The move to the consideration of distance in the plane or in three-space provides a first opportunity to use some of the principles for transfer in planning instruction.

An analysis of the mathematics of distance not restricted to a single line reveals four major problems needing consideration:

1. How can the lengths of segments on two different lines be compared?
2. How can the number that is the length of a broken line segment be found?
3. How can the shortest distance between any two points be found?
4. How can the length of a curve be found?

Each of these problems has a relationship to distance in a one-dimensional setting—a relationship that provides an opportunity to use the principles of transfer.

Problem 1 requires that a relationship be found connecting the measure of distance on line *l* and line *k* (see fig. 2.6). The mathematical solution is to develop the congruence property so that it holds in higher dimensional space as well as on the line. Thus, a common attribute for the structure of measure on a line and measure in the plane or in three-space has been identified. It can serve as a basis for transfer-oriented lesson plans and activities for young children. Also, the types of activities that helped children attain the idea of congruence on a line can be paralleled with comparable activities in higher dimensional space. Parenthetically, it should be noted that making the distinction between congruence of segments on a single line and congruence in a plane hardly seems real to adults because it is so obvious. But as demonstrated in Piaget's seriation interviews, it is not all that obvious to children.

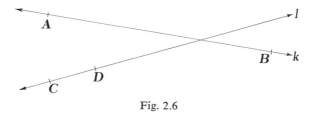

Fig. 2.6

Problem 2, finding the length of a broken line segment (fig. 2.7), will enable children to find the perimeter of polygonal figures. The mathematical solution of the problem involves mapping the segments carefully onto

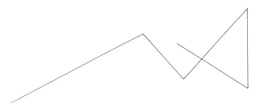

Fig. 2.7

a straight line so that the endpoints of the mapped segments match up neatly, as shown in figure 2.8.

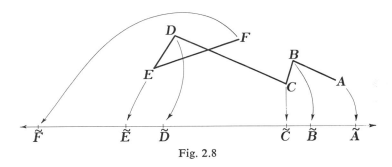

Fig. 2.8

Analysis reveals two characteristics common to the single-line system and this problem in the plane, both of which are important factors in planning teaching for transfer. First, the congruence property, as discussed in the previous problem, plays a significant role. Second, and most significant in this problem, the property of additivity is extended to line segments not on the same line. That is, for a pair of line segments not on the same line, the length of the two segments together must be the same as the sum of their lengths taken separately. One of the principles of transfer states the need to be very careful to establish distinctions or differences relative to the attributes for transfer. For the line, there is a property of betweenness: for length AB + length BC to equal length AC on the line, B is necessarily between A and C so that the segments do not overlap. Overlapping is still a problem if more than one point is common to the two segments. This distinguishing characteristic for the additivity property in the one-line setting and for the higher-dimensional setting is critical. Thus, in an instructional approach for transfer from the single line, the prior learning setting should emphasize the nonoverlapping of the segments. The instruction for the second setting should emphasize that the condition has been lifted. Teachers should create a variety of activities where children examine broken lines and add the lengths of the component segments. They might begin working with broken lines constructed from strips of tape on the floor, transferring lengths (congruent segments) to a number line, and adding the lengths on the number line. However, it is most important that the problem of overlap be included in the activity.

The third problem can be examined from two different vantage points. First, an analysis of the problem again reflects on the property of additivity. The betweenness characteristic of the additivity property has once again been relieved for the purpose of finding the shortest distance between two points. In figure 2.9, $d(A,C) < d(A,B) + d(B,C)$ applies for any point B not on line AC. In teaching for transfer, the emphasis should be on building the idea that given any three points in space, A, B, and C, then $d(A,C) \leq d(A,B) + d(B,C)$. This attribute of the distance function now contains the attribute for the single line case.

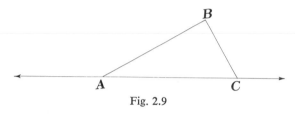

Fig. 2.9

The solution of the final problem, determining the length of a curve, is a step beyond most school mathematics. Even though the solution depends on limit processes, the additivity property, extended to allow for adding an infinite number of segments, does play an integral role in the solution. Consequently, transfer is again an important aspect of the solution.

Area

Area, like distance, is characterized by a function. Many of the properties of the distance function are paralleled by properties of the area function. These properties are among the most important common attributes forming the basis for teaching for transfer. And these very properties are the primitive subconcepts the child needs in order to have control of area for the practical purposes of everyday measurement. Too often slighted by a too rapid progress to formula-based computation, the child needs considerable experience with the primitive subconcepts in order to develop an intuitive feel for area. After the child has developed an intuition for area, the problems for establishing the formulas for the familiar squares, rectangles, triangles, trapezoids, and circles are treated more readily.

The domain of the area function will initially be limited, like the distance domain, in order to simplify the analysis and parallel more accurately the child's encounters with area problems. The domain is limited to the set of all polygonal regions. A polygonal region, such as that displayed in figure 2.10, is a closed, broken line segment together with its interior. A

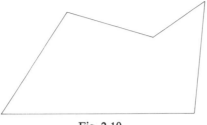

Fig. 2.10

polygonal region may be partitioned into a finite number of triangles by connecting vertices of the region. The set of all polygonal regions is labeled *R,* and the elements of the set are labeled *r.* The area function *A* associates with each polygonal *r* a single, positive, real number. We write

1(*a*). $A:R \rightarrow \Re^+$, or

1(*b*). $A(r) \in \Re^+$.

This definition of the area function is quite similar to the distance function. In each case, the range element is a unique positive number. The commonalities that help the teacher plan instruction for transfer are found within the properties that characterize the functions, which follow.

2. The first property of note is additivity: If r_1 and $r_2 \in R$ and r_1 and r_2 share only points of their boundaries, then $A(r_1 \cup r_2) = A(r_1) + A(r_2)$. The additivity property of area provides a means of associating numbers with combinations of two or more elements of the domain. The area of the union of two regions is the same as the sum of the areas of the two regions. It is important to note, for the sake of designing instruction for transfer, that the nonoverlapping of the regions serves exactly the same purpose for the area function that betweenness served for the distance function.

3. The area function also possesses a defining property of congruence: If r_1 and $r_2 \in R$ and $r_1 \cong r_2$, then $A(r_1) = A(r_2)$. Two congruent regions have identical areas. One of the principles for transfer stated that it was important to design learning activities that differentiate between similar, but not identical, situations in which the child would be tempted to transfer when it would be inappropriate. For distance, if we had equal distances, it could be inferred that the domain elements were equal. For area, this does not hold; the reasoning is only one way. Thus, the child must experience activities specially designed to establish this characteristic of the area function in order to differentiate it from the comparable characteristic of the distance function.

The distance function was strengthened by imposing a coordinate system on the line. The coordinate system was built by the iteration of a unit segment. In order to associate a number with any region in a comparable fashion, the idea of a unit area needs to be established. Used in conjunction with the two properties previously identified, the unit area property is the basis for the tiling, or covering, approach to area. It is usually stated in terms of a square with a side of length 1.

4. The unit area property: If a square, S, has a side of length 1, then $A(S) = 1$. The child's preliminary experiences with area are usually based on covering a figure with unit areas, using the additivity property to provide a base for counting, and thus arriving at the single real number that is the area. The intermingling of several subconcepts characterizes the encounters most learners have with area. Taken together, these concepts are needed to characterize the area structure or system of measure. However, in the interests of transfer of instruction, each of the primitive subconcepts must be progressively differentiated or considered on its own.

The unit property, although particularly suited for children's preliminary encounters with area, does lead to difficulty. The fundamental idea of covering is natural and simple—it allows the child to use counting skills in acquiring intuitions about area. But it is when the child reaches the stage of needing to deal with incommensurability that the counting or tiling basis of area presents difficulty. Instructional sequences are typically designed to evade this difficulty because resolving it takes considerable mathematical maturity. The evasion takes the form of designing an equivalent approach:

4'. (alternate). If a rectangle r has a base of length b and an altitude of length a, then $A(r) = ba$. The usual formulas for polygonal regions are then immediately accessible. This alternative to the unit property plays the same role as imposing the structure of the real numbers on the number line through the coordinate function. The assignment of a unit is implicit in the assignment of lengths to the sides of a rectangle. It ties the measure systems for length and area together.

When the alternate multiplicative concept is used, instruction for the area subconcepts tends to emphasize the range values of the function rather than the characteristics of the function. Classroom talk stresses the outcomes of computations or the answers to area problems rather than the properties of the area function. Consequently, the middle or junior high school teacher concerned with transfer should pay particular attention to finding problems and instructional activities that focus on properties of the function in terms of both the domain and the range of the function. For example, the teacher may want to have students work with the comparison property for the area function.

5. Comparison property: If $r_1 \subset r_2$, then $A(r_1) < A(r_2)$. The instructional activities can be designed to stress the tie between the domain and the range elements; if one polygon is contained within another, then the

area of the former is less than the area of the other. Parenthetically we ask, Is this another property that can serve as a base for designing teaching for transfer? Does a comparable property for length exist?

The analyses of the distance and area measure systems have indicated five properties that can serve as a strong base for applying the six principles of teaching for transfer. They are summarized in table 2.1.

TABLE 2.1

	Distance	Area
1. Function	Distance is a function yielding a single, positive number for each domain element.	Area is a function yielding a single, positive number for each domain element.
2. Additivity	The union of two nonoverlapping segments has the same measure as the sum of the measures of the segments.	The union of two nonoverlapping regions has the same measure as the sum of the measures of the regions.
3. Congruence	Congruent segments have the same measures.	Congruent regions have the same measures.
4. Comparison	Segment $A \subset$ segment B means $d(A) < d(B)$.	Region $r_1 \subset$ region r_2 means $A(r_1) < A(r_2)$.
5. Unit	Coordinatizing the line provides the unit needed to assign a measure to a segment.	A unit is needed to assign a measure to a region. It is usually derived from a related distance function.

A final note concerning the area function and its properties: As with the distance function, we began by examining the area function with the domain restricted to the simplest cases, namely, where the domain elements were polygonal regions. Ultimately, the type of region to which the area function applies must be extended to regions bounded by curves and to regions that are not limited to being within one plane but that are curved surfaces in higher dimensional space. These problems are similar to those encountered when distance was extended from a domain of a single line. It should be noted that the area properties with the incorporation of limit arguments are precisely what are needed to develop the area under a curve in integral calculus.

Volume

The mathematical structure for the measure of volume shares many characteristics of the measure structures for distance and for area. It provides an even better base for transfer because children have had experiences with two structures sharing common attributes. An analysis of volume measure indicates that it is a function possessing the additivity, congruence, comparison, and unit properties summarized in table 2.1. Experience also indicates that a major problem children have with volume is quite similar to a major problem they encounter when learning area concepts—a problem created when premature stress is placed on using formulas and the related computation. Children need experiences directed toward establishing the primitive subconcepts for the volume function in the same way and for the same reasons that they did for the area function. Thus, activities based on using unit-sized blocks to build and count volumes the way area units were used as coverings for polygonal regions can strengthen the learner's intuitions for the subconcepts of additivity, congruence, comparison, and unit.

Volume, however, does present unique problems and difficulties differentiating it from area and length. In the classroom setting, these problems and difficulties arise when the learner must bring together volume concepts for solid measure and for liquid measure. Parallelepipeds and prisms provide a good setting for the unit-filling and unit-counting approach. Additivity, congruence, and comparison seem natural concepts for most learners. With liquid measure, the concepts of additivity, congruence, and comparison seem natural for the mature learner. Most experiences of children with volume measuring initially separate solid and liquid measure systems perhaps partly because this has been the natural thing to do within the English system of measure with quarts, pints, and gallons appearing so different from cubic inches, feet, and yards. This is not a problem within the metric system. But the teacher must recognize that the visual stimuli for liquid measurement and solid measurement differ. The perceptual setting for volume is much more complex. Direct comparisons, such as placing one line segment beside another, are not possible when comparing volumes. Even though the properties—additivity, congruence, and comparison—are the same, the teacher must design instruction for transfer in such a way as to help learners pull the ideas together into a single volume structure.

The volume function is characterized further by the Cavalieri principle. This principle helps the student incorporate both liquid and solid volume measurement into the same cognitive schemata, even though that is not its complete function mathematically. This primitive concept for volume asserts, informally, that although a solid is distorted by the parallel displacement of points in planes parallel to the base of the figure, the range value of its volume remains constant. One can imagine deforming a stack of playing cards by pushing against the edges; one does not change the volume of the deck (see fig. 2.11). This concept is easier for children to

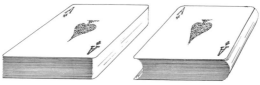

Fig. 2.11

apprehend than to state. It provides a mathematical means of computing the volume of a prism or a cylinder by simply multiplying the height by a cross-sectional area.

Angle Measure

The angle measure function has many properties in common with the measure functions for distance, area, and volume. Since distance and area properties are perhaps more obvious to children, an emphasis on the design of learning activities for transfer is critical for instruction in angle measure. The source of difficulty for angle measure is threefold. First, there is no way of talking about angle measure that is as simple and straightforward as that for length and area. "Length" is the name for the attribute of a line segment, and "area" is a specific symbol for an attribute of a region. But short of the complex mouthful *angularity,* we have no single word for angle measure that is comparable. Moreover, an ambiguity frequently arises when we talk of an angle as a point set, the union of two rays with common endpoints, and the angle as a measure. Frequently, learners encounter statements like "The angle is 45°." The learner must decide from the context whether the word *angle* is used for the angle itself or for the measure of the angle. This type of confusing task is not characteristic of distance, area, or volume.

Second, historically there have been several alternative means of defining angle and its measure, ranging from Ptolemy's half-chords to the present systems used in textbooks. Most of these are quite natural. The young child attempting to find the measure of an $\angle ABC$ (see fig. 2.12) by finding the length k of a segment perpendicular to ray AB through a point some fixed distance from point A has developed a natural approach to the measure of angles. The way that a carpenter uses his square to measure angles for cutting rafters is testimony to the naturalness and efficiency of this means of measuring angles. Whereas for length, area, and volume the

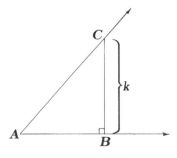

Fig. 2.12

approach to measure seems logical and natural to children, a degree of arbitrariness enters into specifying the measure units and approach for angle.

Finally, the observational tool for measuring angles, the protractor, is more complex to use than a ruler. The child must acquire skills in placing the protractor so that a ray and its endpoint are aligned exactly right; otherwise, the measurement will be incorrect. Also, going immediately to reading the degree scale sometimes interferes with the learner's acquiring an understanding of the role of the unit domain element in angle measure. This is further compounded by the necessity of distinguishing in advanced mathematics between angles of measure module 360; an angle of 65° is to be perceived as different from an angle of 425°.

Angle measure has the following properties in common with distance, area, and volume. First, it is a function, with the critical difference being that the range must be handled in such a manner as to build in a modulo system. Second, the congruence property holds: congruent angles have the same measure. Third, additivity holds for nonoverlapping adjacent angles with a common vertex. Fourth, a comparison property allows one to determine from the measure of the angles which angle is larger. Finally, a unit measure for angles is needed to assure a system sufficiently powerful to solve the types of problems encountered by the learner. Thus, the analysis of the angle measure function reveals attributes that can provide a basis for designing instruction for transfer.

Indirect Measure

Many of the commonly used systems of measure other than length, area, volume, and angularity share a characteristic that provides learners with difficulties. These difficulties are perceptual in nature and stem from the indirectness of the determination of the range values of the function. For example, range values for temperature are determined by reading the length of a calibrated column of mercury or reading a dial. Now this characteristic is a step removed from the qualities of the object the learner should associate with temperature. The qualities of warmth, coldness, heat, and the like are quite different from the length of a column of mercury or the position of a hand on a dial. Mass suffers the same confounding perceptual factor. The standard equilibrium apparatus for determining mass shows the learner a wavering needle indicating whether balance has been achieved when standard masses are placed in the pan opposite the object (fig. 2.13). When a line segment is measured, the ruler is contiguous, but when mass is measured, the needle of the balance may not be at all close to the object being weighed. Also, the needle does not indicate a number as the ruler does; that must be determined by counting the standard masses.

Fig. 2.13

For many of the measures of interest to scientists and of use to children as they learn about, and try to control, their environment, the indirectness of the relation of the underlying perceptual base to the mathematical model and to the scale is unavoidable. The indirectness of this relation introduces a factor of complexity into the learning processes. The teacher cannot rely on directly observable features of the entity being measured to provide the backbone of the instructional design.

The indirectness or perceptual complexity is of three types. First, the unit of measure for length, area, volume, and angularity looks like the object being measured, but a reading of a thermometer does not look like temperature or heat. Second, unlike a ruler or a protractor, the measurement tool does not copy the unit—the wavering needle on a balanced scale seemingly has no relation to the units of mass. Finally, many measurement devices magnify the senses—the seismograph some distance from the epicenter of an earthquake records quivers and shakes of the earth that are imperceptible even to a sensitive human. Each of these perceptual complexities contributes to children's not being able to use their sensory perception as a base for learning about indirect measure systems. This means that the teacher must rely heavily on transfer as a basic instructional strategy.

Some measure systems are derived. For example, the measurement of velocity is in terms of the composition of the measure functions of time and distance. Thus, derived scales also interfere with the students' direct use of sensory data to form measure concepts. However, these problems can be alleviated to some extent by designing learning activities in which all but one of the measured attributes are held constant while observing the effect of the one changing attribute on the measure system. In this way children can use their senses to learn about characteristics of the derived scale.

Although most mathematical instruction is not concerned with indirect systems of measure, many of the learner's encounters in real life and in science are in terms of indirect measure. The functional character of these scales provides the base for understanding that is common to the measure systems more typical of the mathematics classroom. Typically, learners deal with indirect measure when they have the mathematical ability to cope with the range elements of the function and infer characteristics of the domain elements of the system.

Transfer Revisited

The learning of measure concepts has been discussed heretofore in terms of two or more systems of measure and the mathematical characteristics and attributes common to those systems. The goal was to identify in the two systems common features that could serve as a base for the transfer of learning. The assumption was made that students could work, think, and learn across systems of measure. The purpose of this section is to consider the learning within a single measure system in terms of transfer.

A system of measure is a mathematical model. It is a mathematical refinement of what was and is observed in manipulating real objects. The function that characterizes a measure system—for example, the distance function—ties together operations with numbers and the analogous, parallel operations based on the manipulations of real objects. The union of segments parallels the addition of numbers because of the distance function. Learners' intuitions about measure are firmly grounded in their understanding of operations and actions within both the domain space and the range space. The two structures, the domain space and the range space, are related by the function characterizing the measure system. The range space is the mathematical model of the domain space.

The process of relating two structures exemplified by measure systems is called a *homomorphism*. It is a function relating operations in one set with operations in another set. One of the first homomorphisms encountered by a child is the counting function C shown in figure 2.14. The characterizing function for each of the measure systems of distance, area, volume, and angularity relates operations or manipulations within the domain structure to operations within the range structure. Additivity helps to character-

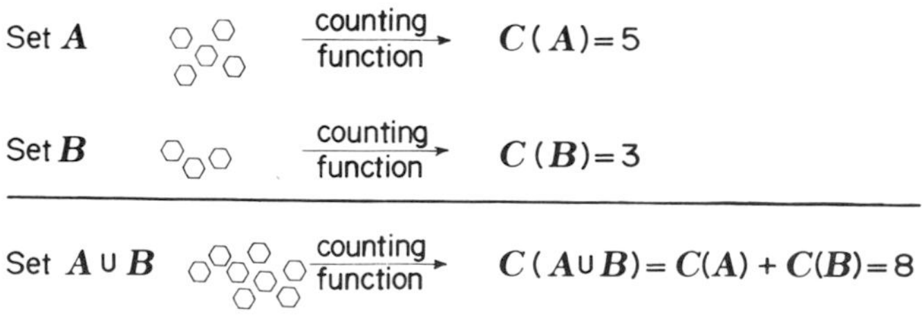

Fig. 2.14

ize each of the measure systems and indicates the homomorphic nature of each measure system.

The learning of a single measure system may be considered as a case of transfer. The learning of the nature of the domain structure provides the base for the transfer of learning to the range structure. Transfer as typically considered by psychologists is not based on two sets of learnings as tightly related as those within a measure system. The homomorphism builds and organizes the base for transfer. The primitive subconcepts identified for each measure system provide the interconnections between the two sets of concepts, principles, and generalizations to be learned. Therefore, instruction designed for transfer within a measure system will stress these primitive subconcepts in order to emphasize the commonalities of the two structures, the range and domain spaces, along with their respective operations and manipulations. This functional connection between the mathematical model and the real-world manipulations is stronger than analogy, common elements, and other mechanisms identified by learning theorists as facilitating transfer.

The homomorphic character of measure and model building is worth stressing in its own right, for it is a concept of significance in modern science and mathematics. It should be observed, however, that the idea is useful in organizing instructional strategies rather than as a concept to be taught formally (and formidably) to children.

Conclusion

Measure is ubiquitous. Measure concepts surround the learner. The learner uses an understanding of measure to quantify and interpret his or her world. This understanding provides the base for instruction for many new mathematical and scientific concepts. As a consequence, the acquisition of an understanding of measure is of fundamental importance to the learner. The learner's extensive and intensive experiences with measure throughout school and nonschool activities do not seem to add up to much power in coping with new measure systems. For a significant number of young people, old learning does not make easier or more efficient the learning of measure concepts in new settings. But analyses of the mathematics of several measure systems reveal that several powerful ideas are common to a large number of measure systems and can provide a base for designing instruction to facilitate transfer.

The question identified at the outset of this essay still holds: Why does transfer of learning take place so infrequently? The hypothesis that has directed the selection and organization of the content of this essay is that instruction needs to be designed to emphasize those common attributes and distinctive characteristics of measure systems. These provide the framework for the design of instruction for transfer. To this end the analyses of the mathematics of measure were directed. Thus, only the first of the six principles for designing instruction to enhance transfer has been accomplished. The teacher must still address the problem of applying the other five principles in the classroom setting.

REFERENCE

National Council of Teachers of Mathematics. *The Growth of Mathematical Ideas, Grades K–12*. Twenty-fourth Yearbook. Washington, D.C.: The Council, 1959.

3

Children's Thinking
in Measurement Situations

Leslie P. Steffe
James J. Hirstein

Mathematics teachers are constantly seeking ways to improve the quality of their students' mathematical experiences. The quality of such an experience is influenced by a number of easily identified factors. In planning the mathematical experiences of young children, for example, the teacher should consider the stages of cognitive development. The proposed content and the methods of presenting that content should also be considered. Finally, the social aspects of the teaching-learning situation should be taken into account. If the teacher could consider each of these factors as being independent, they would not pose the problems they do. The difficulty is that any change in one factor may profoundly affect the others. These interactions can be displayed in schematic form (fig. 3.1).

Although the central purpose here is to consider children's thinking in measurement situations, it should be made clear that different arrangements of content, methods, and social factors will almost certainly affect

This essay was prepared as part of the activities of the Georgia Center for the Study of Learning and Teaching Mathematics, under grant no. PES7418491, National Science Foundation. The opinions expressed herein do not necessarily reflect the position or policy of NSF.

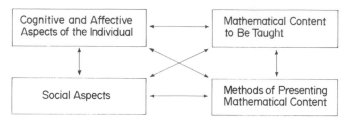

Fig. 3.1. Interaction of factors involved in the mathematical experiences of children

the way a child thinks about the situations. For example, consider a situation of content—a child is asked which of two wooden sticks is longer. If the sticks were physically present, the child could place them side by side and make the comparison in a purely perceptual way. But suppose that instead of two sticks, two line segments were drawn on a blackboard in such a way as to preclude a valid visual comparison. A comparison of the lines would require the child to reason transitively, that is, a string or a stick or some other object would have to be used in making the comparison. Since the lines cannot be brought together, the child must use *reason* to make the comparison; perception alone is not sufficient.

In considering the way a child thinks about measurement as a conceptual system, the mathematics teacher should ask the following questions: How does the child perceive objects to be measured? Can the child use reason in making comparisons among objects? Does the child have a notion of units in measurement? Can the child use and understand various formulas associated with measurement? Does the child know how to use the measurement function?

Length

Comparisons in the domain of length measurement functions are either direct or indirect. Indirect length comparisons involve comparing the lengths of two objects by using a third object. Suppose Helen wishes to compare the heights of two towers, one on each of two tables, by using a rod. Perhaps she compares her rod to the first tower and finds the tower longer than the rod. She then compares her rod to the second tower and finds the rod longer than that tower. Now, to compare the lengths of the two towers, she must reason transitively: since the first tower is longer than the rod and the rod is longer than the second tower, then the first tower must be longer (or higher) than the second one. Transitive reasoning would also be necessary if both towers were the same length as the rod. Substitutive reasoning could be used if the rod were the same length as one tower but either longer or shorter than the other tower.

Direct length comparisons are quite easy for children of school age because they depend on purely perceptual information. Indirect length comparisons, however, require mathematical reasoning and are more difficult. As described, the direct and indirect comparisons are made *in the presence* of physical objects. Therefore reasoning can, and does, occur in the presence of physical objects. But one goal of instruction in measurement is to have children make comparisons *in their heads*—in other words, to reason in the absence of physical bodies as well as in their presence.

Spatial representation is involved in comparing two rods mentally. For example, Warren is blindfolded and given a spatial configuration made from wire. He is to feel the configuration and imagine what it looks like. Obviously, he cannot visually perceive the configuration; it must be reconstructed at the level of thought through tactile exploration. But spatial representation is not simply imagining or reproducing a copy of some physical configuration as a camera records reality. Rather, spatial representation truly involves the *reconstruction* of the physical configuration in thought.

Spatial representation, then, involves mental action. In fact, spatial representation is said to be the internalized and symbolic expression of mental spatial action (Laurendeau and Pinard 1970, p. 14). In dealing with the "shorter than" relation, for example, children must perform the mental action of arranging symbolic segments in order. The fact that they can reason transitively is a result of their ability to order a series of three sticks mentally and symbolically. In other words, they know that there is a subsegment of the second stick the same length as the first stick and that there are two subsegments of the third the same length as the first and second (fig. 3.2). Not only do spatial relations have to be constructed at the level of thought, but so do spatial objects, such as segments, lines, points, planes, planar regions, and so forth. The spatial development of the segment is considered first.

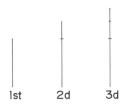

Fig. 3.2

The child's conception of a segment

A procedure used to study the ability of a child to represent a line was introduced by Piaget and Inhelder (1963) and taken up later by Laurendeau and Pinard (1970). Although these investigators all claimed to be studying the development of the projective straight line in the child, their procedures only required the child to construct a row, which is more like a line segment. Since Laurendeau and Pinard described their procedures in more detail than Piaget and Inhelder, major reference will be made to the work of the former.

The specific tasks set for the child by Laurendeau and Pinard involved two pieces of plywood, one rectangular and one circular, each with two small toy houses attached and marking the endpoints of rows of lampposts to be constructed by the child. The child was given eight simulated lampposts and instructed to make a row between the houses with the lampposts. Six tasks were administered (fig. 3.3). Limited in scope as they may be, these tasks illustrate quite nicely the distinction between perception and representation. It is very easy for a child to recognize a segment, discriminate a segment from other curves, and even make a drawing of a segment. However, it is not so easy for a child to make a segment from one place to another if the visual background of the region contains distracting elements. In tasks 1 and 2, the successful child had only to construct a line parallel to the sides of the board on which the two houses were mounted. Tasks 3 and 6, however, provided no such orientation. In order to complete these tasks, the child had to ignore perceptions and depend on a mental image or representation of a line.

Laurendeau and Pinard tested 700 children between the ages of two and ten years. Their findings show that most of these children, even the

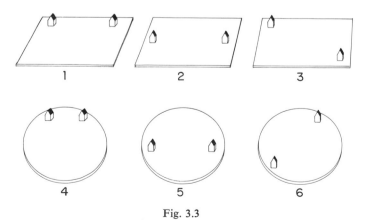

Fig. 3.3

preschoolers, were able to construct a straight row of lampposts when other straight lines were present in the visual field to help orient them. Their data indicate that a lag of several years exists between the ability to arrange lampposts parallel to the sides and the ability to arrange them at an angle to the sides. If the six tasks outlined are critical in determining a child's representation of a line, one would conclude that at least half the children below grade 4 lack the ability to construct a line at the level of representation and are unable to do the tasks that depend on such a representation.

Even though a majority of six-year-old children are not capable of constructing a segment at the level of representation, they are capable of sophisticated perceptual processes concerning geometrical figures—they may need, though, the support of diagrams, drawings, or physical models. Children able to reason at the level of representation become less dependent on such diagrammatic or physical models.

Spatial representation must be considered as a mental ability fundamental to mathematics instruction. Although not teachable in the same way that mathematical concepts are teachable, spatial representation is used by children in processing information of a spatial character. It should be clear that all children of school age are capable of certain kinds of spatial representation, although flexible reasoning is more characteristic of some than of others. One way in which the mature, flexible reasoners can be distinguished from the immature ones is by their display of reasoning ability where measurement functions are involved. Since the transitive and substitutive properties are an integral part of such reasoning and make indirect comparisons possible, it is important that they be considered.

Transitive and substitutive properties

The transitive property has been central in the study of the growth of logical reasoning in children (Smedslund 1963; Braine 1959), but complete agreement on the age at which one might expect children to manifest a capability to engage in transitive reasoning is lacking. Piaget reports that transitive reasoning is evident in some children at about five or six but that only 50 percent are capable of handling transitivity of length relations at about age eight. Using nonverbal testing procedures, Braine (1959, p. 39) concluded that transitivity of length relations appears at least two years before the ages that Piaget reported. But Smedslund, using procedures involving verbalizations, confirmed Piaget's findings. Smedslund's and Braine's findings were clearly contradictory.

If transitive reasoning makes its appearance as early as Braine suggests, then one could introduce indirect measurement problems at almost any grade level and assume that at least 50 percent of the children could handle the problems successfully. In an attempt to shed light on this issue and on the role of learning experiences in the acquisition of relational reasoning, a series of studies was conducted. Carey and Steffe conducted the first of this series using 20 four-year-old and 34 five-year-old children. These children were all Caucasian, ranged in social class from high to low (with a majority in the middle), and had a group average IQ of 113. A test of transitive reasoning was given to these children, as in the Braine and Smedslund studies, but which differed in that the children were given a concentrated and rather sustained learning program involving transitive relations before being tested. Seven instructional days devoted to direct length comparisons for the children were followed by eight instructional days of experiences in indirect length comparisons. The focus of the instruction was on concept formation rather than problem solving. Two tests of transitivity were administered, one after the first seven instructional days and one after all fifteen instructional days had been completed. Little change was noted in the scores for the four-year-olds; only four children showed any evidence of transitive reasoning at all. However, about 30 percent of the five-year-olds could handle transitivity after intense training on the problem. If Braine's contention that transitive reasoning is latent in at least 50 percent of five-year-old children is correct, the training procedures should have activated the reasoning process in that ratio at least. It might be added that even though the transitive property was not easily acquired by the children in the sample, they quite readily developed the ability of making direct length comparisons after instruction.

To check the results of their first study, Steffe and Carey (1972) conducted a similar one using an all-kindergarten sample. The children were selected from two private kindergartens in Athens, Georgia, which served a middle-to-upper-class population. Forty-eight children were involved in an instructional program of twelve successive instructional sessions.

Before the sessions began, a transitive reasoning test was administered, and 33 percent of the children showed evidence of reasoning transitively. The same test administered after the instruction period indicated that 72 percent of the children exhibited this ability. Twenty-two children changed from showing no evidence of transitive reasoning to exhibiting some evidence of transitive reasoning. For some unexplained reason, five children who showed evidence of transitivity in the first administration of the test did not do so in the second.

These results are more consistent with what one would expect from Braine's findings. One must remember, however, that the children in the sample were from the higher socioeconomic levels and had parents able and willing to pay their tuition.

Another noteworthy study in the area of logical reasoning was conducted by Almy and her associates (1971). The intent of the study was to determine whether children would develop more advanced logical thinking capabilities in second grade if they had had prescribed instruction in mathematics and science in kindergarten. Almy (1971) reported that "when the performances of the children who had prescribed lessons beginning in kindergarten are compared with performances of the group about whose kindergarten experience information is available but who did not have the prescribed lessons, we see that the latter do as well as their counterparts who had the lessons (p. 235)." One of the variables included by Almy was transitivity. Of 629 second-grade children involved, 161 displayed the ability to reason transitively. Almy's results and those of Carey and Steffe (1968) and Steffe and Carey (1972) were clearly different.

Owens (1972) conducted an investigation that dealt with the acquisition of transitive reasoning in a population of kindergarten and first-grade children from a low-income, predominantly black, urban community. The children were randomly assigned to one of two treatment groups. The full-treatment group received seventeen instructional sessions on the following topics: direct length and set comparisons (longer than, shorter than, same length; more than, fewer than, same number as), the conservation of matching relations, and the transitivity of matching relations. The partial-treatment group received instruction only on direct length and set comparisons.

The results of only those tests related to transitivity are included in this discussion. After instruction, a test of transitivity of matching relations, a test of transitivity of length relations, and a transitivity problem were administered to all the children. The transitivity problem was included because the test of transitivity of matching relations was purely an achievement test and was not problem oriented. The transitivity of length relations test was included to determine if any transfer of the learning of transitivity of matching relations to the learning of transitivity of length relations had occurred.

On the direct achievement test for transitivity of matching relations, there was no difference between the kindergarten and the first-grade performances. The full-treatment group, however, significantly outperformed the partial-treatment group. The mean score for the full-treatment group was 58 percent and for the partial-treatment group 36 percent—a significant difference. Scores for the full-treatment group were higher than those reported by Almy (1971) and by Carey and Steffe (1968), but somewhat lower than those reported by Steffe and Carey (1972).

There was, however, no significant difference between mean scores on the transitivity of length relations test for the full- and partial-treatment groups. Moreover, the full-treatment group did no better on the transitivity problem than the partial-treatment group. Twelve of the first-grade children could solve the transitivity problem, but only three of the kindergarten children could do it. From the results of Owens's study, it appears that one may train children to *act* as if they can cope with transitivity, but one should not expect any transfer, even when the problems or relations are closely related.

All the data above applied to children aged four, five, and six years. Smedslund (1963) tested children ranging in age from four to nine years for transitivity of length relations. Lamb and Steffe (1975) confirmed his findings, observing that children's ability to use the transitivity and substitution properties of matching and length relations sharply increased between seven and eight years of age—the mean score increased from 44 to 58 percent. The mean scores for the nine- and ten-year-olds were 72 and 64 percent, respectively, indicating that this increase levels off, at least up to ten years of age.

The unit of linear measurement [1]

A conception of a unit of linear measurement logically depends on a capability to engage in transitive reasoning. As a concept, the unit of linear measurement is developmental, just as relational reasoning is. Therefore it is enlightening to describe children's spontaneous reasoning in relational contexts, since this reasoning develops synchronously with a concept of units of measurement.

Again, imagine two towers whose heights are to be compared. Children aged four and five years make their comparisons *visually;* they do not think to move one tower next to the other. Children in the age range of five to seven years use what is called *manual* transfer; they bring the towers close together. The comparison is still visual, but with the towers closer together it can be made more accurately. Children in this age range begin spontaneously to use their bodies to compare the two towers. They will span the height of one tower with their hands or arms to transfer it to the other, a mode of behavior called *body* transfer. Children thus begin to use a third object to compare the heights of the towers, but it is still done on an intuitive level.

Children aged eight years and older will use a middle or third rod in making their comparison. Six-to-seven-year-olds can, with prompting, use a middle rod longer than both towers to compare the towers but not one that is shorter than both, since that would depend on a concept of a unit they have not yet acquired.

1. The theory and data for this section are taken from Piaget, Inhelder, and Szeminska (1960).

It is important to use the spontaneous behavior of children described above in programs designed to train children to make direct and indirect comparisons. Such programs do not necessarily improve children's ability in making indirect comparisons, but they do improve the children's ability to make direct comparison based on manual transfer.

With the advent of the ability to subdivide segments, the unit of measurement intervenes in the transitive reasoning of children. From the foregoing discussion, it is apparent that one should not expect children to be capable of subdividing a segment into subsegments at the level of representation until they possess the notion of a line at the level of representation. What is meant by subdividing a segment at the level of representation is the ability to conceive of a segment as being partitioned into subsegments connected at the endpoints (fig. 3.4). If all the subsegments are congruent, then any one of them might be used to represent a unit.

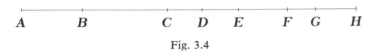

$$A \qquad B \qquad\qquad C \quad D \quad E \qquad F \quad G \quad H$$

Fig. 3.4

The ability of children to subdivide a segment was studied by Piaget, Inhelder, and Szeminska (1960) in the following way. Two strings were each drawn taut between nails, with a movable bead on each string. The strings were denoted by A_1C_1 and A_2C_2 (see problem 1 below). The children were given the following general directions (p. 129): "The bead is a tram traveling along its track. My tram is going this far (moving the first bead from A_1 to B_1). I want you to make your bead do a journey which is just as long as mine, one the same length, etc. Now you see I'm doing this journey (A_1B_1), how far will I have to go on your track (A_2C_2) to do a journey the same length?" These directions were followed for six problems, the first three of which are presented here. The child had an unmarked ruler to use.

Problem 1. Here A_1C_1 and A_2C_2 were aligned parallel, each 30 cm in length and arranged as follows. A_1B_1 was shorter than the child's ruler.

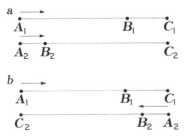

Problem 2. Here A_1C_1 and A_2C_2 were aligned parallel, each 30 cm in length but staggered 10 cm as shown. A_1B_1 was shorter than the child's ruler.

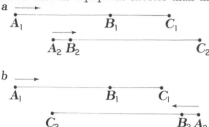

Problem 3. This was the same as problem 2 except that A_1B_1 exceeded the length of the child's ruler and the strings were lengthened to 50 cm.

Certain stages in the growth of measuring behavior were identified by the six problems. Up to approximately seven years of age, children could do problem 1a but none of the others. These children had no difficulty in seeing where the bead had to be placed as long as the points of departure were the same, the paths were parallel, and the beads traveled in the same direction. However, when the points of departure were at opposite ends (so that the direction of travel was opposite) or staggered, then these children failed. In short, children at first think of the distance traveled only in terms of the point of arrival, ignoring the the point of departure. Children progressively improve and finally reach a point where they reason by using what is called *unit iteration,* the repeated application of a unit on the object to be measured. In linear measurement, this involves a change of the position of the unit and a subdivision of the segment being measured.

Children who cannot do all the problems either have difficulty with unit iteration or concentrate on the subdivisions of the segments and ignore the changes of positions of the units, or else they do not conceive of subdivision or change of position at all but merely coordinate the endpoints. For them the function of measurement is not generalized and is often used to explain the results of intuitive procedures.

For the children who can do all the problems, the primary purpose of measurement is to locate the positions of the bead, using visual estimates only when such estimates are appropriate. Subdividing the line and changing the position of the unit are completely synthesized. Although not enough additional information on measurement behavior is available to give percentages of successful performance at various ages, as was done for the representation of a line, some data (Bailey 1973) do indicate the difficulty children experience in measurement tasks.

Bailey administered three measurement tasks to forty first graders, forty second graders, and forty third graders with mean ages of 7.1 years, 7.9

years, and 9.2 years, respectively. The main object of the study was to determine the ability of children at these age levels to compare the lengths of two broken polygonal paths when no formal instruction had been given to the children on the problems. One of the three problems follows, with a diagram of the materials shown.

Problem 1. Here six red sticks each of length 4⅝ inches made up a red path, and six sticks each of length 4¾ inches made up a green path. Three sticks of the lengths indicated were given to the child. The child was asked, "Is the red path the same length as the green path? Use these sticks to help you find out."

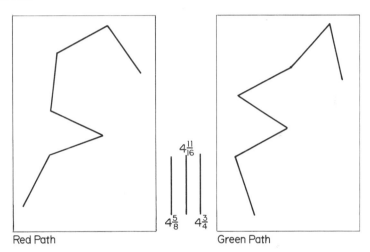

Red Path   Green Path

The child was shown how to use the sticks if they were not used properly. In fact, every subpath of each path was measured by the experimenter with the appropriate stick, if necessary. To solve the problem, the child had to know that every subpath of the red path was the same length, that every subpath of the green path was the same length, that any one subpath of the red path was shorter than any one subpath of the green path, and that there were six such subpaths in each path. The experimenter, if necessary, performed part or all of the physical manipulations, giving the child the correct information while doing so, to insure that each child had a chance to acquire all information necessary to solve the problems. The child had to reason that all the parts of the red path were the same length, that all the parts of the green path were the same length, and that there were six of each; so, because each red part is shorter than each of the green parts, the red path must be shorter than the green path.

Two other, similar problems were presented. In the second problem, seven red sticks were in the red path, six green sticks were in the green path, and *all* the sticks were 4⅝ inches in length. Here, the red path was longer than the green path. In the third problem, the two paths were the same length because there were six 5-inch red sticks in the red path and six 5-inch green sticks in the green path. All paths were staggered as shown in the problem.

Piaget, Inhelder, and Szeminska (1960) found that at least 50 percent of children aged between eight and eight years six months had attained operational measurement. It was expected that the third graders in the study reported here would solve the problems, since their mean age was 9.2 years. However, this was not so—only four third graders solved the problems correctly and were able to explain why their solution was correct using both the number of paths and the lengths of the subpaths.

Area and Volume

Spatial representation is important not only for length measurement but also for area and volume measurement. Diagrams and physical models of spatial regions are heavily used in the teaching of area and volume. Presenting children with a diagram or a physical model of a spatial region in no way insures that they can make a faithful mental reconstruction of the diagram or the model.

The child's conception of a region

Although Piaget and his associates (1960) claimed they were studying the conservation of area, it could just as accurately be described as the conservation of planar regions. The methods of study were quite simple. One method consisted of presenting the child with a rectangular region divided into subregions and then rearranging the subregions to ascertain if the child considered the region to occupy as much space as before. The procedure was to show two regions made up of, say, six square regions and then transform one of the regions as in figure 3.5.

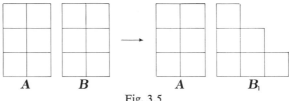

Fig. 3.5

The child was asked whether B_1, the transformed region, had "the same amount of room" (i.e., as much area) as A. In order to check the results, the experimenter took ninety-six wooden cubes, which covered the 2 × 3 rectangular regions (A and B) exactly, and asked the child whether the cubes would cover B_1 as well.

In the second method, the child was shown two congruent rectangular regions; then a piece was cut off from one region and moved to another part of the same region. The question asked was, "Are these the same size?" or "Is there the same amount of room?"

The results obtained from these experiments parallel those obtained in the section on subdividing a straight line. Up to about six or seven years of age, there is absolutely no conservation of a planar region. Later, children answer some questions correctly, but their successes are the result of trial and error. Finally, at about the age of seven or eight, the child can resolve the problems posed. These children insist that the transformation did not alter the space—"its the same space; you just cut it!"—and that the wooden cubes would also cover the transformed rectangle.

The three-dimensional region presents difficulties not encountered in one or two dimensions. Again, Piaget and his coinvestigators (1960) claimed they were studying the conservation of volume. Regardless, their study sheds light on a child's concept of spatial regions. The investigators identified three aspects of what they called the *volume concept:* internal volume, occupied space, and displacement volume. This breakdown has been used in subsequent studies by Lunzer (1960) and Lovell and Ogilvie (1961[a]).

Internal volume may be considered as the number of units inside the boundary of a spatial region. The child was shown a block 4 cm in height with a base 3 cm by 3 cm. The child was then induced by the experimenter to build a house on a base 2 cm by 2 cm, 2 cm by 3 cm, 1 cm by 2 cm, 1 cm by 1 cm, or 3 cm by 4 cm having the same amount of room as the first house. The question was whether some rearrangement of the cubes would produce "the same amount of room." The task is similar

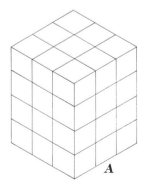 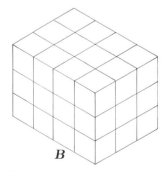

Fig. 3.6

to the problem with planar regions. It is difficult to complete without
having a concept of volume as a certain number of unit cubes. However,
it is possible to complete the task using one-to-one correspondence,
thereby producing a spatial region, such as *B* in figure 3.6, without knowing the number of unit cubes.

For volume as occupied space and for displacement volume, some
cubes were placed in water. The question was whether rearranging the
cubes would change the level of the water. Occupied space is the
amount of room taken up by the total region; displacement volume is
the amount of water displaced by the region. The distinction between the
two points of view must be reflected by the questions asked. Lovell and
Ogilvie (1961[a]) used the following procedures to distinguish them:

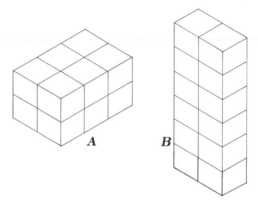

1. *Occupied space:* Before a can is filled with water, block *A* is placed inside. The can is emptied and block *B* is placed inside. The child is asked
   if as much water could go in the can as when block *A* was in it.
2. *Displacement volume:* Block *A* is lowered carefully into a full can of
   water, spilling some of the water. The child is asked to compare the
   amount spilled to that which would be spilled if block *B* were lowered
   into the can.

The responses of children on these tasks differ from the results
described earlier for length and area. Children at first exhibit no conservation behavior. Then comparisons are made, but the comparisons are
usually based on a single dimension. For example, the child would build
tower *B* the same height as tower *A* but disregard the bases. In the
age range of seven to eleven years, children are capable of solving
problems involving the conservation of internal volume, but they are
unable to conserve occupied volume or displacement volume, an ability
that comes somewhat later.

The preceding data concerning internal, occupied, and displacement
volume are some of the best data available concerning the child's concep-

tion of spatial regions. If the definition of volume as a number is adhered to, then tests of the conservation of volume would explicitly involve units of volume measurement and the number of such units. The tasks on volume used by Piaget and his associates (1960) explicitly involved units of volume but only implicitly involved the number of such units. Whereas this comment does not invalidate the experiments, it suggests that the experiments in fact do deal with the child's conception of a spatial region. Any alteration of a given region certainly changes the region in the sense of producing a region of different shape. However, a child who can build a tower of dimensions different from a given one of the same number of unit cubes is capable of partitioning a given region, such as a $3 \times 3 \times 4$ block, into smaller rectangular parallelepiped regions and rearranging these rectangular parallelepipeds mentally to produce another spatial region containing the same space. For instance, the child might produce a series of cubes 36 cm long, end to end. It is in this sense that Piaget and his coworkers were studying the child's conception of spatial regions. They provided tasks that would help them determine the child's ability to decompose a given region into subregions and then mentally rearrange those to produce a new arrangement of the subregions. The experiments on occupied and displacement volume, in addition to testing internal volume, test the ability of a child to conceive of a given region in the context of its surroundings, a most critical aspect of a conception of spatial regions.

The unit of area measurement

As with length measurement, the idea of a unit for the area of planar regions begins with subdivision. The child must be able to conceive of a region as a union of subregions connected at the boundaries, as in figure 3.7. The entire region R is the union of the subregions R_1, R_2, R_3, R_4, R_5, and R_6. If the subregions all have the same area (but not necessarily the same shape), then any one represents a unit, and the area of region R is six units. The difficulties with this conception result from two requirements. First, the child must accept that the whole region is composed of the subregions and realize that the component parts can be

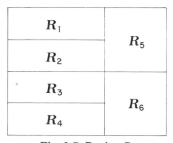

Fig. 3.7. Region R

disassembled and then reassembled to give the original region. The second requirement is that the component parts must have the same area. Piaget and his associates (1960) have studied the development of this concept with a task involving the sharing of a circular cake. Their conclusion is that of the two requirements, the parts-whole aspect (the first) takes longer to develop and that once it is developed, the matter of equal-sized parts develops relatively easily.

Once the concept of a unit is present, the next step is the iteration of the units to assign a number to a given region, which is true measurement. Most studies have employed regions that can be covered with tiles, either squares or triangular halves of squares, although the mathematical notion of unit is actually more general. Children's ability to iterate area units has been studied by two methods. In one, the child is given enough identical tiles to cover both regions to be compared. In the other, only a few tiles are available, so that the subdivisions cannot be made by completely covering the regions; the child, however, is given a pencil and is allowed to mark the regions. The conclusions are that (1) a sequence of abilities occurs that exactly parallels the development of length measurement and (2) competence is attained for both length and area at about the same age (Piaget, Inhelder, and Szeminska 1960).

Beilin and Franklin (1962) have disagreed with the second of Piaget's findings, that is, the simultaneous acquisition of the notions of a unit of length and a unit of area. They used algorithmic procedures to try to train six- and eight-year-old children to use unit iteration in both length and area situations. The older children were successful in both contexts, but the younger children could learn only the length procedure. One difficulty appeared to be the inability of the children to consider two directions at the same time, which was required in the iteration of area units.

In an attempt to extend Piaget's findings, Wagman (1975) devised an area-unit iteration task employing units of two different sizes on the same regions (fig. 3.8). The child was given enough square tiles (unit 1) to cover all three regions. After the child had covered each region, the tiles were

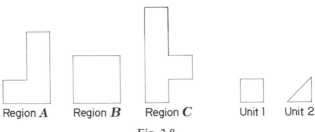

Region *A*   Region *B*   Region *C*   Unit 1   Unit 2

Fig. 3.8

stacked beside the region to aid in memory. Then the child was given a number of half-squares (unit 2) and asked to find how many triangular tiles would cover each region. Whether actual covering was done or not was left to the child to decide.

The responses to the triangular-unit task fell into four categories: (1) the child had to tile each region in order to answer; (2) the child concluded the response for region B from its equality with A (known from tiling with the square units); (3) the child developed a rule after tiling one or two regions and then responded on the basis of the rule; and (4) the child compared the square and the triangular units and then based a response on the relative size of the units. Wagman found that only 3 out of 25 eight-year-olds needed to tile all the regions to answer the question.

These results may be influenced by the fact that little perceptual conflict is caused by the units. In an unpublished pilot study in 1974, Hirstein asked children to cover a rectangle with two different sets of rectangular cards (fig. 3.9). The cards were not congruent, but it took six cards to cover the rectangle. Not until age ten did any children conclude that a red card and a blue card had the same area. Younger children always used only a single dimension to compare the areas.

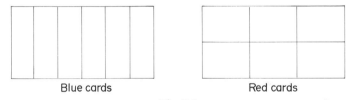

Blue cards   Red cards

Fig. 3.9

The conflicting results in the Wagman and the Hirstein studies can be readily explained. In the Wagman study, the children may look at the area units as arithmetical units—as just something to count. This was entirely possible because they were all congruent within a task. It has been found that the arithmetical unit is an earlier development then the area unit (Piaget 1952). In the Hirstein study, the additional feature of noncongruent units was involved, making it much more difficult for the child to know that the units were of equal area. The conclusion that the two units were of equal areas had to be drawn entirely from the knowledge that it took six of each to cover the same rectangular region.

Piaget and his associates (1960) found that children do not think of the area of a rectangle in terms of the product of the lengths of the sides until approximately twelve to thirteen years of age. Children know very early in school that $2 \times 3 = 6$. They can also be taught that a two-by-three rectangle can be conceived of as two rows of square regions with three squares in each row, and since $2 \times 3 = 6$, then 6 is the area of the rectangle. Of course, it is more general that the lengths of the sides will not be whole units. That the area is still the product of the lengths of the sides demands a conceptualization that goes beyond the simple situation where the sides are an integral number of units in length. Even with whole numbers, it is an open question whether the child conceptualizes the area as the product of the lengths of the sides. A test for such a conceptualization would demand that one start with the length of the sides and then ask for the area, and little information is available on the problem other than that generated by Piaget and his associates (1960).

The unit of volume measurement

The child's conception of the unit of volume measurement has not been studied very thoroughly. The results discussed here are somewhat of an extrapolation from experiments having a different focus. Piaget's group (1960, pp. 354–89) discuss, in theory at least, the connection among the unit of volume, the kind of volume (interior, occupied, and displacement), and the multiplication of length, width, and height to find the volume of a rectangular solid. Younger children, when asked to reproduce a pictured tower, frequently built a configuration containing only sides. These children did not visualize the parts of the surface units that were not visible. Older children could construct or produce another tower of different dimensions, but when asked to calculate its volume, they reverted to behaviors common in younger children. As they grow older, children, when asked to calculate interior volume, either equate volume with a given number of unit cubes, where the number is taken from the boundary surfaces, or equate it with the number of unit cubes it takes to surround the model.

Finally, beginning at about twelve or thirteen years of age, children (1) develop the ability to view volume as interior volume, as occupied space, or as its complement, displacement volume; (2) develop the notion of a unit of volume; and (3) associate the product of the lengths of the sides with volume. In these children, the unit is coordinated with the lengths of the sides, and displacement volume and volume as occupied space are coordinated with internal volume. The unit of volume, then, gains new significance in that the child can use it to tell how much space is occupied by a spatial region, what the possible dimensions of a rectangular solid are, and that the product of the lengths of the sides of a rectangular solid remains constant as the lengths of the sides are varied through some rearrangement of unit cubes.

Another study of interest was conducted by Lunzer (1960). He used children ranging in age from six to fourteen and found no evidence that any of them thought of volume in terms of what is surrounded by boundary surfaces. (This negative finding does not mean that Piaget and his group did not find children who did regard volume as consisting of boundary surfaces, but rather it indicates that children who regard volume as such are quite limited in number.) Lunzer did verify that the notion that a spatial region is composed of units that can be altered without changing the internal volume was understood by only the older children. The ages at which Lunzer found conservation of displacement volume were approximately the same as those Piaget and his associates reported. However, Lunzer's interpretation of his data was quite different, since he gave a more prominent role to mathematical instruction in school. He indicated that children in school have few experiences with displacement volume. This, in addition to the fact that the conservation of displacement volume depends on the conservation of the volume of water that is displaced, seems to account for the late acquisition of the conservation of displacement volume. Lunzer also thinks that few children would spontaneously acquire the knowledge that the volume of a rectangular solid can be found by multiplying together the lengths of the appropriate sides. He contends that this notion is the result of school instruction and not of maturation.

It is not possible at this time to know whose interpretation is correct. Lovell and Ogilvie (1961[a]) replicated some of the work of the Piaget group, but their study was not designed to shed light on the issues raised by Lunzer. They found that 40 percent of children aged about eight years conserved volume as occupied space compared to 80 percent of children aged about eleven years. For the older children, however, their conception of physical volume was strongly influenced by the weight of the object, the depth to which it was immersed, and the size of the containers. Lovell and Ogilvie suggest that it *may* be possible for school experiences to hasten the acquisition of volume.

Carpenter (1971) administered thirteen tasks, only two of which are discussed here. In one task, the child was shown two identical glasses containing equal amounts of water and was asked to compare the amounts in the two glasses. Adjustments were made, based on the response, so that the child would agree that the two glasses contained equal amounts. The water in each glass was poured into two opaque containers, one taller but narrower than the other (fig. 3.10), using two visibly different units of measurement of such sizes that one glass of water measured three units and the other five units. The children were then asked to compare the amounts of water in the two opaque containers. A second task was exactly like the first except that the smaller unit of measurement *appeared* to be the same size as the larger. The tasks were administered to a total of sixty-one children in the first and second grades. Of the sixty-one children, twenty-six concluded there would still be the same amount of water in the two opaque containers after the distinguishably different units were used, whereas only twenty came to this conclusion when indistinguishably different units were used. The children gave interesting reasons for both the correct and the incorrect responses. Fourteen of the twenty-six children who responded correctly to the first task reasoned that the amount of water in the two opaque containers was the same because the amount of water was the same in the two original containers. These children successfully ignored the fact that three repetitions of the larger unit were used versus five of the smaller. Nine of the remaining twelve successful children reasoned that the same amount of water was in the two opaque containers because one unit was smaller than the other, and so more of them would be used. On the second task (indistinguishable units), eighteen of the twenty children referred to the previous state of the water in the two original containers. None of these twenty children compensated for the number of units by pointing out the difference in size.

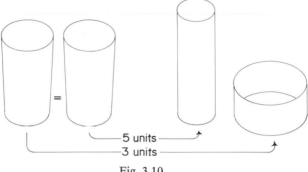

Fig. 3.10

Approximately 60 percent of the children tested were unsuccessful on both tasks. All but three of the incorrect responses were based on the one glass measuring more units than the other. These children understood neither the role of the unit nor its arbitrary nature. Carpenter's results agree with those of the other investigators: the conservation of internal volume is only gradually elaborated during the first three or four years in school. Four distinct levels in the development of conservation and measurement were suggested by Carpenter:

1. The child responds on the basis of a single dominant dimension, whether visual or numerical.
2. The child is capable of changing from visual to numerical considerations but in a given task tends to remain focused on one dimension. Children in this level seem to ignore the present state and refer to former states, stating, for example, "They were the same before." This marks a transition to higher levels of thought concerning conservation and measurement.

3. The child gains the flexibility to consider several conditions of a quantity simultaneously and chooses the condition that provides a rational basis for comparison.

4. The child can use the information from both the initial and the final states of the water to discover the correct relation between unit size and the number of units, a process essential for rational measurement.

Teachers must not conclude that all tasks involving length and area measurement are solvable by children eight years of age or older. The type of experience engaged in will vary from task to task. The volume concept confirms this statement, since the notion of internal volume occurs prior to that of either occupied or displacement volume. The situation is not completely chaotic, however—one could expect that length and area tasks involving congruent units could be handled by children aged eight years or older. Tasks with unequal units or noncongruent units are more difficult. Tasks incorporating both length and number are quite difficult, and one would not expect much success before eight or nine years of age. The notion of internal volume appears to develop at about the same time as notions of length and area, but those of displacement and occupied volume are later developments.

Weight

Piaget (1941) has considered the development of weight conservation in conjunction with volume and the quantity of matter. The experimental procedure is the same for all three. Two balls of clay are judged by the child to be the same for all three attributes. One of the balls is then rolled into a "sausage" or sliced into pieces, and the child is asked whether the attribute in question is the same. The general findings are that (1) somewhere between eight and ten years of age, children begin to believe that the two balls of clay have the same amount of substance, but (2) not until about age ten or twelve do children believe the weight of the balls remains the same, and (3) usually not until after age twelve are they convinced the two take up the same amount of space. These conclusions of Piaget have been substantiated in replications by Elkind (1961) and Lovell and Ogilvie (1961[b]).

Physical attributes such as weight can be developed only through experiences with objects—for example, comparing weights and measuring with equivalent units. Using devices like the pan balance to give the child operational uses of weight concepts seems to enhance this development, although Smedslund (1961) found that subjects who were *trained* to conserve weight abandoned this reasoning more readily than subjects who acquired it naturally: he found that if, when the "sausage" was rolled out, a small piece was removed without the subject's knowledge, only those subjects who had become operational *naturally* concluded that some must have been removed.

Time

Piaget (1969; 1970) has also undertaken a number of experiments to ascertain the development of the concept of time in children. The concept of time is mentally constructed by coordinating the speed at which various actions occur. The primitive notions of speed are derived from a moving object passing another object. If object A passes object B, then object A must be moving faster than object B. This operational definition of *faster* is based on information that is strictly ordinal. If object B is ahead of object A, then the "passing" occurs whenever object A is ahead of object B. These data depend, not on how far behind one object is or how long it takes to pass, but only on the order of the objects.

Evidence that children accept speed by a "passing" argument is given by the following experiment. Two dolls are made to move on tracks through two tunnels of different lengths. The dolls begin at the same instant and emerge from the tunnels at the same instant. Four- and five-year-olds will conclude, even though they have no trouble identifying the longer tunnel, that the dolls travel at the same speed through the tunnels because they start and stop together. When the tunnels are removed and the one doll can be seen to overtake the other, the children decide that one doll moves faster than the other. However, when the tunnels are replaced and the dolls are sent through again, many of the children return to their original belief that the dolls are moving at the same speed.

Once children have become aware of the preliminary notions of speed, they can begin to coordinate actions with the speed at which they occur. It is these coordinations that lead to the construction of the concept of time. Two major obstacles must be overcome in this construction. The first is the idea of the sequential occurrence of events, the realization that event A occurred before event B. The second is the concept of a length of time, or duration, as in the time it takes for a glass of water to fill. The problems are best illustrated by children's reactions in experimental situations.

The first experiment shows how a child's perception of a sequence of events can be affected by conflicting information. Two objects are set to move along parallel tracks (fig. 3.11). The objects start at the same time. Object 1 moves from A_1 to D_1 in the same interval that object 2 moves from A_2 to B_2. Then object 1 stops at D_1, but object 2 continues at its same speed to C_2, where it stops. Young children maintain that object 2 stopped first because it did not get as far, despite careful attention to insure that the wording of the questions is not confusing the time with the distance.

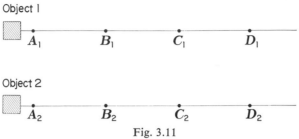

Fig. 3.11

A variation of the previous experiment shows the problem involved in conceptualizing the length of a time interval. On the one hand, the two objects start at A_1 and A_2 at the same time but move at different speeds. They stop at the same instant, but object 1 is at B_1 and object 2 is at C_2. Here it is usually accepted that the objects started at the same time but did not stop at the same time because one went farther. On the other hand, if the objects start at B_1 and A_2 at the same time and arrive simultaneously at C_1 and C_2 and stop, the situation is reversed. Here it is agreed that the objects stopped at the same time but that one must have started earlier because it had farther to go. These children are not accounting for the speed of the objects but instead are substantially equating the distance traveled with the time it took, *in spite of the fact that the events occurred in precisely the same time interval.* Piaget (1969) reports that a substantial number of six-to-seven-year-old children make these errors. In a replication of Piaget's experiment, Lovell and Slater (1960) found that only 43 percent of nine-to-ten-year-olds concluded that the objects "traveled for the same time." The percentage for younger groups was even less.

Units of temporal measurement

The measurement of time differs from the other physical domains in one significant aspect: intervals of time are not directly comparable. Once past, two intervals of time cannot be retrieved to be started simultaneously for comparison. In addition, a succession of equal units can be determined only by repeating an event for which each occurrence must be assumed to take the same amount of time. The constancy of speed at which an event takes place is the basis for the construction of the time concept. We are all aware that a minute spent doing something enjoyable seems to go faster than a minute of drudgery, although because we believe in the constancy of the time it takes the second hand of a watch to make one revolution, we are certain that the minutes really take the same amount of time. We have learned, not to rely on what the interval *seems* to be, but rather to place our confidence in a more predictable device. This confidence is precisely what is lacking in the child during the process of constructing a concept of time.

Piaget (1969) has shown that young children adapt the velocity to account for perceptual differences in time intervals. A mark is made on the bottom half of a sandglass, and children are given a job to do until the sand reaches the mark, perhaps transferring marbles from one container to another. At first they are asked to transfer the marbles slowly. Then, with the sand started over again, they are asked to transfer the marbles more quickly. On both tasks they must stop when the sand reaches the mark. Children up to about seven years of age believe the sand moves more quickly when the marbles are transferred quickly. Since the time seems to pass faster while they are working quickly, they evidently believe the sand is falling faster. Only later do they realize that the sand falls at a fixed rate independent of the speed of transfer. The use of an external device to compare intervals will not be meaningful until this independence is accepted.

Because the concept of a constant velocity must be accepted before comparisons of time intervals can be made, Piaget concluded that speed is a more fundamental concept than time. However, even the acceptance of the constant velocity of a single object is not sufficient to construct the time concept. Children may still be confused when comparing the same interval with two different devices, such as a watch and a sandglass. The devices may seem to move at different speeds, causing children to predict that one device would allow them more time, even though the intervals have been demonstrated to be simultaneous. Not until the age of eight or nine do children believe that simultaneous intervals take the same amount of time. Until children are able to compare intervals determined by different actions, we cannot expect them to comprehend the measurement of time.

REFERENCES

Almy, Millie. "Longitudinal Studies Related to the Classroom." In *Piagetian Cognitive-Development Research and Mathematical Education,* edited by Myron F. Rosskopf, Leslie P. Steffe, and Stanley Taback. Washington, D.C.: National Council of Teachers of Mathematics, 1971.

Bailey, T. G., Jr. "On the Measurement of Polygonal Paths by Young Children." Doctoral dissertation, University of Georgia, 1973.

Beilin, Harry, and I. C. Franklin. "Logical Operations in Area and Length Measurement: Age and Training Effects." *Child Development* 33 (1962):607–18.

Braine, M. D. S. *The Ontogeny of Certain Logical Operations: Piaget's Formulation Examined by Nonverbal Methods.* Whole no. 475. Psychological Monographs: General and Applied, vol. 73, no. 5. n.p., 1959.

Carey, Russell L., and Leslie P. Steffe. "An Investigation in the Learning of Equivalence and Order Relations by Four- and Five-Year-Old Children." Research paper no. 17. Georgia Research and Development Center in Educational Stimulation, December 1968.

Carpenter, Thomas P. "The Role of Equivalence and Order Relations in the Development and Coordination of the Concepts of Unit Size and Number of Units in Selected Conservation-Type Measurement Problems." Wisconsin Research and Development Center for Cognitive Learning. Technical report no. 178. Madison: University of Wisconsin, 1971.

Elkind, David. "Children's Discovery of the Conservation of Mass, Weight, and Volume: Piagetian Replication Study II." *Journal of Genetic Psychology* 98 (1961):219–27.

Lamb, Charles E., and Leslie P. Steffe. "A Study of Children's Ability to Use Inference Patterns with Matching and Length Relations." Paper read at the 53d Annual Meeting of the National Council of Teachers of Mathematics, April 1975. Mimeographed.

Laurendeau, Monique, and Adrien Pinard. *The Development of the Concept of Space in the Child.* New York: International Universities Press, 1970.

Lovell, Kenneth, and E. Ogilvie (a). "The Growth of the Concept of Volume in Junior School Children." *Journal of Child Psychology and Psychiatry* 2 (1961): 118–26.

——— (b). "A Study of the Conservation of Weight in the Junior School Child." *British Journal of Educational Psychology* 31 (1961): 138–44.

Lovell, Kenneth, and A. Slater. "The Growth of the Concept of Time: A Comparative Study." *Journal of Child Psychology and Psychiatry* 1 (1960):179–90.

Lunzer, E. A. "Some Points of Piagetian Theory in the Light of Experimental Criticism." *Journal of Child Psychology and Psychiatry* 1 (1960):191–202.

Owens, Douglas T. "The Effects of Selected Experiences on the Ability of Disadvantaged Kindergarten and First-Grade Children to Use Properties of Equivalence and Order Relations." Doctoral dissertation, University of Georgia, 1972.

Piaget, Jean. *The Child's Conception of Number.* London: Routledge & Kegan Paul, 1952.

———. *The Child's Conception of Time.* London: Routledge & Kegan Paul, 1969.

———. *Genetic Epistemology.* New York: W. W. Norton & Co., 1970.

Piaget, Jean, and Barbel Inhelder. *The Child's Conception of Space.* London: Routledge & Kegan Paul, 1963.

———. *Le développement des quantités chez l'enfant.* Paris: Delachaux & Niestle, 1941.

Piaget, Jean, Barbel Inhelder, and Alina Szeminska. *The Child's Conception of Geometry.* London: Routledge & Kegan Paul, 1960.

Smedslund, Jan. "The Acquisition of Conservation of Substance and Weight in Children: III. Extinction of Conservation of Weight Acquired 'Normally' and by Means of Empirical Controls on a Balance." *Scandinavian Journal of Psychology* 2 (1961):85–87.

———. "Development of Concrete Transitivity of Length in Children." *Child Development* 34 (1963):389–405.

Steffe, Leslie P., and Russell L. Carey. "An Investigation in the Learning of Relational Properties by Kindergarten Children." Paper presented at the annual meeting of the American Educational Research Association, April 1972. Mimeographed.

Wagman, H. S. "A Study of the Child's Conception of Area Measure." In *Children's Mathematical Concepts: Six Piagetian Studies in Mathematics Education,* edited by Myron F. Rosskopf. New York: Teachers College Press, 1975.

4

Teaching Measurement to Elementary School Children

James E. Inskeep, Jr.

M easurement in elementary school mathematics holds a unique position. It isn't really mathematics in the sense of a deductive discipline, nor is it a strictly cognitive area in which facts and knowledge play a prime role. Measurement joins activity with cognition and geometry with arithmetic. To teach measurement we should be concerned with more than just the system of measures. We must teach a "doing" kind of mathematics—a practical, socially useful skill. Activity gives meaning to the measuring skills, makes the resultant learning personally satisfying to the child, and begins the development of a process that will be used throughout life.

But what of metric measurement? Many people feel that teaching the International System of Units (SI) automatically simplifies the teaching of measurement. There has been a rush for metric materials, and rightly so. However, obtaining materials and developing a new vocabulary will not in themselves insure that measurement will be taught effectively—that is, so that it becomes an important part of the learner. Although SI is far less confusing than customary U.S. measures, neither system requires much cognition.

What must teachers know in order to teach measurement effectively? They must know something about how children learn it and about how measurement ideas and activities should be presented. They should have some idea of the overall objectives of teaching measurement and a feeling for sequencing activities to foster these objectives. Accordingly, this essay will attempt to (1) present some fundamental principles for the effective teaching of measurement to children, (2) list a set of overall objectives (goals) for the measurement program in the elementary school (K–8), and (3) give examples of teaching techniques and approaches that correspond to the objectives.

Principles for Teaching Measurement

To develop principles for teaching measurement, we must know something about the learning process. We could approach the problem from an empirical/theoretical point of view, applying the research efforts of psychologists to the task of measurement. Or we could analyze the act of measuring and develop some subskills or sublearnings leading to a well-developed skill. Since the empirical/theoretical approach is well documented elsewhere (see the works of Piaget and his interpreters), we shall discuss only the analytical approach. When we seek to reduce the measurement process to a sequence of substages, we also establish goals for our curriculum. This analysis need be neither exhaustive nor totally sequential to give guidelines for our instruction. It should, however, help us teach with a systematic completeness the preskills necessary for the development of adequate measurement skill.

How does a child learn to measure? If we analyze the process, we find it is actually the merging of important sensory and perceptive skills with the cognitive aspects of geometry and arithmetic. It also involves the affective area and provides a child with the opportunity to feel a sense of accomplishment and to appreciate the basic usefulness of our measuring system. The process proceeds sequentially from perception to comparison and then to the application of a standard measure (or referent). Spencer and Brydegaard suggest principles for measurement as a mathematical behavioral pattern (1966, chap. 2). Some of their ideas, with modification, will be noted in the next section.

Measurement as perception

Measurement starts with a perception of what is to be measured. Explaining the markings on a thermometer to a child without first developing some sensation and perception of what is being measured is just another drill in scale reading. A child's height, for example, gives meaning to length, whereas weight does not.

As adults, we take for granted that children perceive as we do. Most children do have some experiences that allow them to develop perception of the world about them. However, this is frequently left to chance and seldom developed in a systematic fashion. A teacher should be willing to expose children to many stimuli and many properties of objects that will eventually be "measured." These activities become fundamental beginnings for learning skill in measurement.

Measurement as comparison

Perception is the beginning of measurement, and comparison follows perception. Having perceived a property of some object, we naturally compare it with other objects having the same property—if one jar holds so much liquid, would not another jar hold a different amount? When we put our hand into a container of water and withdraw it, we experience a change in sensation caused by the evaporation of the liquid. What do we have to compare with this experience? Does it feel the same way after we take a bath? Is the feeling anything like the one we experience when we open the refrigerator door or come close to an air-conditioning duct?

Comparisons of sensations are quite natural. Comparisons of objects that can be placed close together are also natural consequences of perceptions. In measuring their height, some children may wish to know how their height compares with that of other children in the room. We may then direct the children to lie down on large sheets of paper and to draw around each other's body shapes and display the outlines in such a manner as to compare them. This activity is done without any number ability. Comparing attributes of objects leads quite logically to the need for a standard that we can apply over and over again. We may now view measurement as the quest for a standard or referent.

Measurement as the quest for a referent

Comparing two things is adequate when we wish to make gross statements of equivalence or nonequivalence: "You are taller than I am." "I am taller than my little sister." These serve well for initial comparisons. They may even serve for logical comparisons with third parties. "If I weigh more than my brother and he weighs more than my little cousin, then I weigh more than my cousin." However, it is soon apparent that this approach to comparison is quite ineffective. We really need some standard of measure, a referent that may be used over and over again and to which we can return at any time. The initial referent we use need not be a standard referent or one that is used throughout the world. For example, parts of the body are readily available referents for measuring length.

Nonstandard referents are useful for comparison, but we do wish to take our children beyond the obvious and teach them about the referents that can be used with more than one person—our standard measures.

Standard measures have at least two important functions. First, they permit one person to communicate a measure to another in an abbreviated, direct manner. Second, they allow accurate and consistent measuring in different geographical areas. When we move from one state to another, we can be sure that the measures that are standard in our state are standard in another state as well. A logical extension of this idea would be to adopt usable standards of measure that communicate the same message in all parts of the world. This leads naturally to the International System of Units (SI), which now fulfills (almost!) this worldwide function.

Measurement as a system

With SI, we have a system of related standard measures and have thereby largely replaced the arbitrariness of local standards. It has taken several hundred years for the system to find widespread acceptance in the world, but the end is in sight.

At this point in our thinking, we have taken the process of measurement through several stages—perception, comparison, the need for a referent, and finally, the need for a system that organizes and systematizes the standard referents. The same process can be applied to the educational experience of children. It suggests a sequence of activities. Children are led from early perceptual experiences to the point where they relate these experiences to other properties and tie them together in a systematic manner. At this final point we may say that a child has learned to measure.

Thus far we have neglected certain nonsequential aspects of measurement—those of the affective domain and those linked to the actual *doing* of measurement. Affective outcomes of measurement and the act of measuring are two principles we must consider.

Measurement as an affective activity

Our work with children in measurement should result in two basic affective outcomes: (1) children should appreciate the role measurement plays in their lives and in society, and (2) children should enjoy being able to measure for themselves.

The importance of measurement in our personal lives and in society is often taken for granted. The scientist knows its importance, and the engineer can't avoid it; but the average citizen sometimes fails to appreciate the role of measurement. Children should learn the important role measurement plays in great scientific/technological breakthroughs. Correlating the

mathematics program with science and social studies aids in this development. Introducing some of the measuring skills in art and physical education is also helpful. But children also need to see measurement as an
important part of their own lives. They need to see that it is important
to measure accurately a particular board for the construction of a tree
house. They need the skill to read a clock so that they will not miss a
favorite television program. Children should be aware of the consequences of sloppy or ineffective measurement in constructive activities. No
child who has inaccurately measured the salt for a simple recipe will soon
forget its implications—especially if he had to eat the product!

Another affective characteristic of the measurement process, and one
more difficult to define, is the satisfaction a child can feel from having
done a good job of measuring. Children should be taught to measure so
well that they develop self-confidence because of it. They should feel that
they have accurately described what they wanted to describe and also
understand any limitations. Teaching children that no continuous measure
is exact must be balanced by giving them adequate experience in reading
instruments and scales. To be able to read a new type of scale is a satisfying accomplishment. How many children can read the water meter?
How many children, as they leave the elementary grades, have learned to
read a meterstick and apply that skill with consistent results to similar
objects? The answers to these and other questions suggest what our commitment to teaching measurement should be.

Measurement as an activity

The importance of doing measurement has already been stressed. Measurement without doing is merely some sort of rote memory or intellectual
exercise. Children can memorize the SI units and even assimilate the customary U.S. units without much effort. However, we want more than the
ability to answer standardized test items! We want children to have experiences in all the basic areas of measurement and to be able to measure
accurately and consistently. Such experiences must be systematically
planned by the teacher and become an integral part of the curriculum.

The Objectives of Measurement

We have dealt with some of the principles underlying the effective teaching of measurement. The basic objectives of teaching measurement in the
elementary school follow:

1. *Children should be able to perceive properties of objects to be
measured.* This objective deals with measurement readiness. Children
should be given activities that lead them to experience such properties as

heat, length, weight, time, area (covering), light, texture, volume (space filling), and others.

2. *Children should be able to compare objects that have a similar property.* The experience of comparison must be provided almost simultaneously with perception. When children are made aware of the properties they perceive as length or weight, they should also be given opportunities to compare many objects having the same properties.

3. *The convenient nonstandard measure should become the first referent of comparison.* Children must see the need for a referent or standard of measure. Normally we use the term *standard* to mean a referent that has national or local applicability. In this instance, we shall seek "standards" which the children themselves develop and that are, at best, applicable only to the immediate classroom. When we agree on whose "foot" we shall use in measuring the length of the room, we are establishing a local standard.

4. *The standard measure should replace the nonstandard measure as an accepted referent.* Children must know the standards that are accepted by most of the people they will meet. Since our own country is at a crossroads in its commitment to SI, this may mean that we shall be operating with two standard systems for some time. When we switch to SI, we should teach it directly as *the* system of comparison. However, we should bear in mind that children will be encountering customary U.S. measures through most of their lifetimes.

5. *All children should understand the system of measurement known as the International System of Units (SI).* Since the need for standard referents can be satisfied by using SI and since we shall probably be involved in mass education for metric conversion, we should teach SI as the basic system. This will pose problems initially, and many teachers will continue to teach the customary U.S. measures. Our customary measures will undoubtedly persist for a long time, but to expedite the changeover to SI units, all children should be given their *primary* instruction in the International System of Units.

6. *Children should have an appreciation for the role of measurement in their own lives and in society at large.* This objective is very broad and, to be effectively described, needs a multitude of behavioral descriptors. (Sanders addresses this issue in the opening essay of this yearbook.) These behaviors should become obvious as teachers work with children and seek to teach them measurement in a context that fosters appreciation.

7. *Children should derive personal satisfaction from the actual process of measuring.* As previously noted, this depends in part on the skill children develop in measuring. Teachers can proceed toward this broad

objective through positive reinforcement and encouragement to try various tasks or projects.

8. *Children should be able to use measuring instruments effectively and accurately.* This deals in part with a motor skill. It involves a large amount of practice and feedback. Accuracy and effectiveness in measurement will not come through mere reading or memorization.

9. *Children should be able to read measuring scales, especially those based on the number line.* The most common scales are based on the number line. Some are variations of it, but they generally deal with a simple number line of equal divisions. Children must learn how to read scales and how to estimate when reading between marks.

10. *Children should understand that all measurement of continuous quantities is approximate.* This may be a corollary of objective 9. However, many activities should stress the possibilities for error in measurement—from both the standpoint of the measurer and the instruments. Do repeated measures of the same object produce the same reading? If not, why not? The answers to these questions tend to support the idea that all continuous measurement is approximate.

11. *Children should be able to apply their knowledge of measurement by constructing or making objects from given measures.* This is an inverse to the measuring activity of previous objectives. Here we start with given measures and expect some construction of an object.

12. *Children should have continuous experience with the doing or activity of measurement from preschool up through the grades.* This objective ties the activity-oriented goals together. No one skill in measuring can be obtained through a single exposure. The affective objectives will not be met by a one-time experiment, no matter how well planned. Measurement must be taught as an integral part of children's environment from the time they enter elementary school to the time they leave it.

In summary, these twelve objectives define a program of measurement that should be sufficient for the elementary school. The total measurement program should be ongoing and continuous, as suggested by objective 12. The topics of objectives 6 through 11 should be continually and persistently emphasized in the program. Finally, the sequence of objectives 1 through 5 should be maintained. In these twelve objectives lies the framework for an effective measurement program.

The Content of Measurement, Grades K–8

The preceding twelve objectives provide guidelines for *how* to teach measurement. The content of an effective measurement program remains to be defined. The elementary school years are a base for (1) an introduction to socially useful measurement and (2) the development of sophisticated measurement in science and other subjects at the secondary and postsecondary levels.

Introducing children to measurement that has social utility is essential to the definition of content for the elementary school years. The following types of measures are basic: (1) length, (2) mass (weight), (3) time, (4) temperature, (5) area, (6) volume, and (7) angles. Our first four objectives should include activity with all the preceding types of measure. In dealing with the other objectives, we may use more than the seven measures listed. These seven, however, should be fully developed through the stages of perception, comparison, and formal measurement. These measures are also basic to those developed in the secondary and postsecondary levels.

Some Activities for Teaching

The twelve objectives provide a framework for the activities that follow. Sequence is vital in dealing with the first four objectives, and some experiences with each of the seven types of measures should be provided. On the other hand, the nonsequential objectives would likely include only samplings from the different measures. If we have covered the seven basic types of measures in our instruction, developed an understanding of measurement, and given children experience with other measures, we shall have done our task in teaching measurement in the elementary school.

A constraint in working with young children should be mentioned. Not all children are ready for certain types of measurement or measuring activity. Going through the first four objectives with young children may be a waste of time if they are not intellectually ready for the experience.

The research of Piaget and his interpreters has supported the idea that there are stages in children's lives when they are unable to understand and comprehend the measuring process *effectively* (Copeland 1974 [a] and [b]). Since each child is different, there is little hope for predicting a grade level at which a class of children should be given the activities. If teachers are unable to deal individually with their children, they need other criteria for developing instructional sequence. The experience of the Hawaii Metric Project provides a simple set of guidelines (King and Whitman 1973). Children were given tasks of the Piagetian type in the pilot stages of the project, and it was felt that most of them did not possess the necessary concepts for understanding measurement until they were about nine years old. Hence, the project developed readiness types of activities for kindergarten through grade 2 and did not emphasize formal measurement until the third grade.

Evidence from the Hawaiian project indicates that the first two objectives (perception and comparison) should probably be handled in the lower grades (K–2). Objective 3 can probably be introduced in the primary grades, most effectively at the third-grade level. Objectives 4 and 5 should probably be developed at or above the third-grade level. Most of the activities dealing with objectives 8, 9, 10, and 11 will be most effective at the upper-grade levels (3–8). Objectives 6, 7, and 12 are applicable at all grade levels.

One word of caution: Since each child develops in an individual pattern (this applies to Piagetian concepts as well as to other aspects of individuality), it is not only wise but essential for teachers to *adapt the program to their own students and use their own judgment on when to begin an activity.*

Objective 1: Perception

To an adult, a child's perception seems obvious and perhaps not even essential. It seems "reasonable" that a child should perceive length or weight. However, we often miss the most obvious. One of the functions of the teacher is to help children perceive measureable attributes. One way to do this is to single out and name aspects of the immediate environment. Another way is to measure objects *for* the children, thus exposing students to the process even before they are able to comprehend it completely. We may also point out to them things that are measureable. We may compare attributes that are measureable but different. Perception is readiness; it is vocabulary building; and it is a foundation for understanding.

Perception of length

The following activities may be used to develop or heighten the perception of length as a measureable attribute of the environment:

1. Capitalize on those occasions when children are measured for cumulative records. When the nurse measures the children, have the class discuss what is being measured. The methods used to find the measure of the height may also be discussed. The class may even wish to construct a "height-measuring device" of its own. Ask the children for ideas on how they might find out how tall each one is. Even if a child cannot read scales or compare numbers, *participating* in the activity will help clarify what is meant by height.

2. Ask the children if they know who lives the farthest from school and how they can tell. Develop the idea (and word) of *distance* by asking the children to come up with ideas on how they might find out who lives the farthest. If possible, use a walking field trip to develop the concept of distance. Make visits to nearby stores, homes, or points of interest the basis for discussion.

Perception of weight

The perception of weight parallels that of length, since both are easily associated with living things. The weight of objects can be sensed directly. Holding two objects and comparing their "feel" allows direct sensory experience. Here are some activities for developing a perception of weight:

1. Use occasions when children are measured for their records to develop the idea of weight. Length and weight represent a child's own characteristics. This is strong motivation when approached properly. Give the children the numbers that represent their weights. Exercise care in comparing children's weights, but if it is done naturally and discreetly, one child may be said to be heavier or lighter than another child.

2. Measure a pet's weight. Let the children hold the pets in their hands to "feel" their weight.

3. Provide a number of objects that vary in volume and weight and ask the children to hold them. Take two of the objects and ask them to guess which is heavier. Repeat with several children, each time asking the class to guess which is heavier. Repeat with several children, each time asking the class to guess which is heavier. Holding things in their hands gives children experience with what gravity does to mass to produce the feeling of weight.

Perception of time

Children do not really understand time and its passage until they reach the upper elementary school grades. Even then, many children have only a meager understanding of the passage of time and almost no conception of historical time. The perception of time as a measurable attribute goes on throughout the elementary school years. Some activities to aid in developing this perception follow.

1. Use every opportunity to develop the idea of descriptors of time: the class comes to school in the *morning;* the lunch period separates the *morning* from the *afternoon;* the children go home in the *afternoon.* Using the language of time should be a continual part of early childhood education.

2. Even though the children may not be able to read or tell time, point out to them the differences in the position of the hands on a clock. Have the children sketch what the hands of the clock look like at a given time, followed by another sketch an hour or so later. If this is difficult to plan in the classroom, put a clockface on the chalkboard and ask a child to put hands on it as the classroom clock indicates. Drawing several clocks with different times will help the children visualize the change the passage of time makes in the clockface.

3. Keep track of days, months, dates, and other calendar events. Birth-dates and special occasions help heighten interest for "reading" dates, days, and months. Point out the change in the year when children return to school after New Year's Day. Although these activities do not neces-sarily directly contribute to children's perception of time, they do prepare them for the vocabulary they will use in expressing it. Passage of time can be noted in the upper grades in terms of days, weeks, and months.

4. Measure a pet's growth and associate it with time. Make simple graphs of the pet's height and weight and relate them to the days (and dates) of each measure. This helps tie the change in the pet's statistics to the change in time. Let the children predict the weight of a pet a week ahead of time for additional experience in noting time intervals. Have the children compare their predictions with the measured values.

5. When the children are able to tell time or read calendars, have them perform various experiments that involve the recording of times and events. Comparing changes over a time interval helps establish the ideas of time. Checking an egg timer against a stopwatch aids in developing a feeling for short time spans. Developing some sense of longer time spans is difficult and complicated by the fact that time seems to run slower or faster depending on the individual's occupation during the interval. Marking dates, making schedules of TV viewing times, and developing planned activities involving time intervals (such as writing a radio or TV script) all contribute to the individual child's perception of time.

Perception of temperature

Children are exposed to variations in temperature early in their experi-ence. They may associate warmth with a season or with a location, such as the fireplace. Some children adjust so well to changes in temperature that they do not seem to notice differences. All children must have some concept of temperature as the measure of heat. Following are some activi-ties designed to develop the perception of temperature.

1. Set out two containers of water, one with sufficient ice in it to make it cooler than the other. Ask a child to put one hand in the cool water and the other in the warm water. Ask what is felt and how it is described. Use words like *cooler, hotter, warmer,* and phrases like "the temperature is higher in this container." A thermometer may be used to measure tempera-tures of both containers of water. Similarly, have children hold their hands over a heating or cooling outlet and compare the sensation of that air with the air in the rest of the room. Sharing and discussing such experiences can lead to sharpened ideas of temperature.

2. Let the children experiment with bits of dull black metal and bits of reflecting metal held in the sun. Which is warmer? Which seems to warm up quicker?

3. Keep a record of the weather on a day-by-day basis and compare the reported temperatures. This helps to familiarize children with the thermometer. Compare and discuss temperature differences across the nation.

4. Ask children to estimate the temperature of a container of water. Let the children compare the results of their perceived feelings with the measured temperature.

Perception of area

The perception of area can be developed from primitive ideas of covering objects. Ordinarily, area is merely a convenient means to communicate how much plane surface can be covered. It can be extended to coverings that are not planar. Some activities for developing a perception of area are given next.

1. Most classrooms have bulletin boards that serve as excellent visual aids. Use these as a challenge for covering. "How much paper will we need to cover the board?" "How many pieces of colored paper will we need for the border outline?"

2. Give the children some closed plane shapes and provide them with sets of square, circular, and rectangular pieces of paper. Ask them to cover each shape with each set of the pieces of paper. Record the results, or let a child in a small group record the group results, or perhaps let individual children keep track of their own results. Talk about what they find out. "Which set works best? Why?"

3. Give the children cylindrical boxes or cans and ask them to determine how much paper it will take to cover the curved surfaces. This activity is particularly appropriate during Christmas or preceding Mother's Day, when they are making small gifts such as pencil holders. The circular bottom of the cylinder may also be covered.

4. Potato printing serves as another opportunity to develop some idea of covering. Cut potatoes in half and have the children make designs in the cut ends and dip in ink. Ask the children to experiment with their potato printer by completely covering a small section of paper without any overlap. Let the children record the number of impressions needed to cover the paper. Perhaps point out that large designs on the "printer" require fewer impressions than small ones. Also, help the child see that a figure that does not tessellate the plane (a disk is a good example) leaves holes that are not covered.

Perception of volume

The perception of volume parallels that of area but is more difficult to grasp. Experiences with volume should involve both liquid and space-filling measures. Some of these activities follow.

1. For an introduction to volume, measure liquids with nonstandard containers. "How many capfuls does it take to fill this plastic bottle? How can we find out?" If convenient, have the children do their own measuring, counting the capfuls and recording them.

2. Boxes and blocks furnish another outlet for experimentation with volume. Classrooms in the lower grades usually have large wooden blocks. Since these should be put away at the end of the day or activity time, ask the children to decide how many of them will go into a given box or on the shelf where they are kept. Their answers to these questions, no matter how unreasonable, serve as openings for the development of the idea of volume.

3. Dienes Multibase Blocks, Cuisenaire rods, or Stern materials may also be used to develop ideas of volume. "How many cubes will fill this container? How many small cubes will it take to construct a large cube?" Answers to these questions and the free play associated with a child's own problem solving serves to develop the perception of volume. Play with large and small blocks affords additional opportunity for pointing out the idea of "space filling."

Perception of angles

The need for angular measure is most evident in the upper grades. Except for telling time, few angular measures are encountered in the lower grades. Some simple activities to introduce the idea of angles are listed here.

1. Children who can tell time and have some knowledge of fractions can use the clockface for an introduction to angles. Provide the class with dittoed sheets of clockfaces, unmarked except for indications of where the numerals go. Ask the class to make a line from the center of one of the clocks to the position where the 12 should be. Have them make a second mark on the same face indicating where the long hand of the clock would be if it pointed to the 6. "If the long hand moves from the 12 to the 6, what fraction of the way around has it traveled?" "What geometric shape is made by the first and second positions of the long hand?" (line segment, "straight" line) Repeat the procedure with the initial location at the 12 and the final location at the 3. "What geometric figure do the two positions make?" (right angle, "corner") "If we consider a turn around the whole face as one turn, how much of a turn does the large hand make when it moves from the 12 to the 3?" Try other starting positions and other angles. Have the children make a record of all the starting and ending positions for an angle that looks like a corner of the paper. Adapt the questions and materials to an activity unit or to an exercise that can be done individually.

2. Use a simple exercise in how to measure angles with a protractor to develop the perception of angular measure. (This lesson also meets other objectives and is best handled in the upper grades.) Show children how to measure some dittoed examples of various angles from 0° to 180° with the protractor. A large chalkboard-sized protractor and a demonstration can enhance this activity.

3. Have children measure angles of various plane figures to refine their perception (and comparison) of various angular measures. Give the children a set of triangles and ask them to measure each angle in each triangle. Record the measures for each triangle. "What is the sum of the angle measures in each of the triangles? Is there any pattern? What are the measures of angles in rectangles? Quadrilaterals? Pentagons?"

Objective 2: Comparison

Comparing objects that share a common measurable attribute is a natural second step to perceiving that attribute. Lessons that involve perception quite often involve the comparison of measures as well. When children begin to talk about something being hotter, longer, or heavier, they have perceived and are comparing the differences in that attribute. The vocabulary of comparison is very helpful in developing this objective: "This angle is greater than that one." "The air from the duct is cooler than the air outside." "My foot is shorter than yours." "Our rat is heavier today than he was yesterday." A few activities to develop the ideas of comparison follow.

Comparison of length

Most adults compare the length of one object to another by determining whether the number associated with its measure is greater or less than a similar number for the other object. In our examples of length comparisons, we wish to develop the kind of feeling for length that is both tactile and visual and does not depend on measuring skill or the ability to read and correctly order numbers.

1. "Which bean plant has grown the most in the last week?" This question can serve as the basis for extended measuring and comparing. "How should we measure the height? Could we use strips of paper or string to show how high each plant is?" Note that when this activity is used in the lower grades, children's readiness for measurement is not fully developed, and some adult supervision may be needed to help the children obtain accurate measures. The children may arrange the strings or paper in some sort of order and in effect make a graph of the results.

2. Draw outlines of the children and place them along a bulletin board or chalkboard. Then ask for comparisons: "Who is the tallest? Shortest?"

Measure the height of several children by having them stand next to the chalkboard and marking their height with a piece of chalk. Write the name of the child beside each mark. Make comparisons by looking at the marks.

Comparison of weight

As in the comparison of length, we do not want children to be hampered by a dependence on a measuring instrument or on an understanding of the number order. Weight is best "felt" through the muscles. When children support an object against the pull of gravity, they begin to understand (perceive) what weight is. Comparison develops when two objects are held and some statement is made about which is heavier (or lighter). The following activities illustrate this idea.

1. Obtain some rocks or other objects whose weights vary. Ask the children to put them in order from the heaviest to the lightest according to how they feel. This sort of activity may be packaged and used as an individual or small-group task (a box, some rocks, and simple directions). Mark down the actual measures of the weights of the stones to facilitate checking to see how well a child has compared the weights. Follow up this activity naturally by measuring the weights on a pan balance and then checking the results. (See activity 3.)

2. Ask children to hold the pets they have in the room. "Which pet is the heaviest?" Have them make a list of the pets showing which is the heaviest and the lightest and post it on the bulletin board or the chalkboard. Some discussion should accompany this exercise, and several children should be involved in holding the animals.

3. Use a simple pan balance. Give children regular assignments and have them check the results they obtained from holding their pets. Let them devise experiments of their own to decide which objects are heaviest and keep simple records of their experiments. Ask children to guess which object is heavier before they try it out on the balance.

Comparison of time

As previously noted, the comparison of time intervals is difficult even for the adult. An hour watching a favorite TV program hardly seems the same as an hour practicing addition facts! When measuring instruments are unavailable, an awareness of the passage of time depends on body rhythms (hunger, heartbeat, sleepiness), periodic nature events (sunset or sunrise, moonrise), or regularly occurring events in our daily lives (mealtime, school beginning and ending, playtime). Where possible, these "available" measures of comparison can be used. The measurement of time with clocks and calendars can be developed and refined if these basic feelings for comparison are regularly developed. The following activities precede, or are concurrent with, measuring time with instruments.

1. Give children a feeling for the passage of a day's time by relating it to reading the calendar and observing natural daily events. "Did the sun shine today?" "What is today on the calendar?" Use these questions daily as you mark a large, prepared calendar. "How many times will the sun rise during a whole week on the calendar?" This and other questions can serve to stimulate informal discussion about the length of a day and a week. Extend the idea to develop comparative concepts of the month, season (or quarter), and year. Upper-grade children need these comparisons; younger children can be exposed to them.

2. "How long is a second? A minute? An hour? Which are longer time intervals?" These questions can be answered in simple experiments or by observations. A pendulum that is a meter long has a period of approximately one second. A person's pulse rate will vary—at rest, it may be from sixty to eighty beats a minute. Measure the pulse rates of the children or let individuals do their own. Also compare pulse rates during rest time with those obtained immediately after recess. Pick one who has a rate of about sixty beats a minute. Ask other children to measure this child's pulse, counting to sixty. This count will "define" approximately one minute. Use a metronome to help children get an idea of rhythm in counting. Compare the swing of a pendulum to the pulse rate to give children an approximation of whether their own rates are more or less than sixty a minute.

3. A unit on telling time may provide an opportunity for children to compare given lengths of time. How may we measure time? Some answers might be obtained through the following: a pendulum; sand running through a hole in a can; water dripping out of a bottle; heartbeat; the movement of the shadow of the sun; bouncing a weight attached to a spring. Compare these means for telling time with standard means to highlight the passage of time and fix intervals in terms of physical events. How often must we turn over a simple egg timer to observe six minutes of silence? Can we make a container with a hole in it that lets the water drip out in exactly one hour? These questions and the quest for their answers help to explain the passage of time and the comparison of time intervals.

Comparison of area

Comparing areas is complicated by the fact that direct observation is misleading. As adults, we may look at a triangle, a circle, and a square and yet be unsure which is the largest or the same in area. We may even make these shapes have the same measure, and yet they appear to be different. The comparison is further complicated for children by the fact that they may not have the cognitive structures available to them to make comparisons, even when the task appears simple to an adult (Copeland 1974 [a], pp. 74–79). The following activities are illustrative and presuppose the learner's involvement.

1.   Construct a series of shapes of varying area measure by using graph paper. Ask children to arrange the shapes in order from least area to most area. After they have ordered them, ask them to count the squares that cover the shapes and check their estimates. Marking the measure on the backs of each of the shapes will provide feedback and make this activity suitable for an individualized packet or module. Compare rectangular, triangular, curved, and irregular shapes to make a series of activities.

2.   Use the geoboard to develop comparison of areas. Give the children a particular set of shapes and ask them which is the largest shape in area. Construct shapes on the geoboard and count the squares needed to make the area measure of the shape.

3.   Trace on graph paper one hand of each of several children. Which hand is the largest in area? Let the children count the squares of the graph-paper to find out whether their estimate was correct. Repeat with feet or shoe soles.

Comparison of volume

One of the classic tests for the conservation of liquid volume is to pour water from one of two equally filled (and shaped) glasses into another of different shape. Children frequently feel that the volume has changed. No amount of *immediate* experimentation seems to convince the children that their responses are incorrect, but experimentation *over time* does help them formulate ideas of conservation. Several activities that help children make comparisons of volume follow.

1.   Obtain several plastic bottles having different capacities. Number the bottles and pose the following questions and instructions as an independent activity: Which bottle holds the most water? Which holds the least? Put them in order from the one that holds the least to the one that holds the most. How can you find out whether you have placed the bottles in the correct order? Use one of the caps to fill each bottle and record the number of capfuls it takes. Were you correct in your order?

2.   Obtain a single, clear plastic, noncylindrical bottle with a cap. Down one side of the bottle paste a strip of paper on which a child can make a pencil mark. (Some kinds of transparent tapes or masking tape will serve the purpose.) Ask the child the following questions: Where on the side of the bottle can we put a mark that shows when it is half full? Mark the point where you think it will be. Now measure the volume of the bottle,

using the cap. How many capfuls fill the bottle? Fill the bottle with half the number you found. Were you correct in estimating where the bottle was half full?

3. The following gamelike activity is designed for two children. Obtain a set of twenty cubes, either some of the commercially produced materials (Cuisenaire, Dienes, Stern) or ordinary blocks that are cubes. In this game each child first participates and then acts as the "teacher" for the other. "Take three blocks. How many different shapes can you make using these blocks?" Both students should work together to find these answers. "Use four blocks. Now try it with five or more. After you have tried this with a set of ten blocks, one of you will play the teacher. That person will make two different shapes, neither of which has more than ten blocks in it, while the other person is not watching. When the two shapes are finished, the teacher asks the partner which is the larger. If the partner guesses correctly, the partner becomes the teacher and gets to do the same thing. If the partner answers incorrectly, the teacher gets another turn." Each time this is done, both children should count the blocks needed to make each of the two shapes. If the partner fails to answer correctly three times in a row, the teacher wins the game. It then becomes the partner's turn to be teacher, make the shapes, and ask the questions. The teacher may try to fool the partner by making each shape with the same number of blocks, whereupon the partner must say, "Both are the same." Vary the game by awarding points for correct answers and let the children keep some records. Set a limit on the number of turns to be taken.

Comparison of angles

Perception and comparison are closely related in all beginning phases of the measurement of angles. Since we expect angle comparison to be done in the upper grades, we may deal with both perception and comparison at the same time. When children are given simple measuring tasks associated with angles, ask them to order them in some fashion, the smallest to the largest. Ask them to estimate sizes, check with the protractor, and repeat the process to develop a sense of comparison of angular measure. (For activities, see the section entitled "Perception of Angles.")

Objective 3: Nonstandard Measures

Although it would seem more efficient to introduce standard measures after we feel that children are ready to measure, a strong argument exists for introducing nonstandard measures in initial, formal measuring activities: nonstandard measures are easily available. Most children can use their hands and feet for simple measuring tasks. Similarly, the covering of a bulletin board need not be expressed in square centimeters—the area is just as easily noted by the number of pieces of construction paper it takes to cover the board.

Nonstandard measures help relate the measuring process to the child's immediate environment. Beginning with meters, liters, and grams may sound good to the adult but may not be significant to the child. Using readily available means to develop measurement skills aids in developing an appreciation for the importance of measurement.

The side effects encountered in problem solving are another reason for beginning with nonstandard measures. A child who has been taught to measure length with a ruler often cannot solve simple measurement problems when the ruler is not available. Teaching children to begin measuring with readily available measures, whether standard or nonstandard, is a step toward improving their practical, problem-solving ability.

Since the process of using nonstandard measures is similar for all sorts of measurement, separate activities will not be listed for each of the seven types noted earlier. Instead, we shall list suggested nonstandard measures to be used in lessons involving actual measuring experiences. The main thrust of this objective is to get children into the measuring act, using some local standard that they can employ in talking about their measurements. There should be no compulsion on the part of the teacher to require the memorization of nonstandard equivalent measures. The focus is on the process. Suggestions for nonstandard measures follow.

Length: Parts of the body (foot, hand, length and width of fingers, tip of finger to nose, elbow to tip of finger); strides (for longer distances); and readily available objects (pencils, edge of the paper, lines on a piece of composition paper, paper clips, erasers, strings of a given length, sticks, straws, toothpicks).

Weight: Bolts; nuts; stones of similar weight (this may be checked on a pan balance); paper clips; small containers filled with water and capped; bags of gravel or sand (also weighed on a pan balance to establish the same weight for each bag); blocks of wood or plastic.

Time: Sand clock; water clock (or bottle with small hole for water to drip from); weight and length of string for pendulum; heartbeat (varies with each child); spring and weight for bouncing a counter.

Temperature: Thermometer (the exception to this list).

Area: Small pieces of paper or cardboard (each piece for a given set should be congruent) in various shapes (triangles, rectangles, squares, hexagons, others); small ceramic tiles; and floor tiles or carpet squares. (Note that nontessellating shapes, such as the disk, can be used to demonstrate the need to cover completely each section of an area, leaving no "holes" in it.)

Volume: Bottle caps; plastic cups and glasses; small plastic containers; congruent cubes, small congruent boxes.

Angles: Corners of cards; arbitrary paper angles; bisected paper angles (fold and cut along the crease formed); small segments of a circle cut in pie-shaped figures (may be cut to any arbitrary unit); georules and compass (to be used in constructions to determine comparative sizes of angles and to construct similar angles).

Objective 4: Standard Measures

After the children have gained experience with nonstandard measures, it is natural to shift to activities using standard measures. Since this will most likely occur in the upper grades, some discussion of the importance and use of the International System of Units (SI) should be developed. One way to introduce SI measures is to relate SI to our decimal currency and numeration system. If the customary U.S. measures are being introduced, they may be related directly to many of the nonstandard measures (e.g., cup, foot).

Good lessons focus on the *doing* of the measurement. The transition from nonstandard to standard measures may be gradual or nearly simultaneous. The following example using length illustrates a series of activities that begins with the development of nonstandard measurement and moves to the use of standard units. Other types of measure could be introduced in a similar fashion.

Ask the class questions that involve finding lengths. "How long is the room? How high is the door handle from the floor? What is the width of the chalkboard? What are the dimensions of the desktops? Books? Boxes of materials?" Divide the class into small groups and ask them to answer each of the questions and to record their answers. Have them measure without rulers, metersticks, tape measures, or other standard measuring instruments. Each group must work together, agreeing on the "standard" they use for each measurement. When the measuring is completed, bring the groups together for discussion. Point out that they have used different standards for their measurement. Discuss the difficulties in communicating the measures of one group to another. Explain that they could use standard measures, which are the same throughout the world. Then introduce the meter, the centimeter, and the millimeter and show their relationships. Then, using tape measures, metersticks, and other metric measuring devices, repeat the exercise with SI measures. Discuss the results. The discussion should turn up the fact that although not everyone obtained exactly the same results (particularly when small units were used or an estimate was needed), their results are much closer than when nonstandard units were used. (This activity lends itself to further discussion about accuracy of measurement.)

The foregoing series of activities may be done over a span of time and not attempted in a single lesson. Start with the development of the non-standard measure, lead into standard usage, and then go to the idea of the accuracy of measurement.

Following are some SI units that warrant use early in formal measurement. Some teachers may wish to teach different prefixes and other items attendant to SI, but these should be the primary units with which the children become familiar:

Length: meter (m), centimeter (cm), millimeter (mm), kilometer (km)

Mass (measure of weight): kilogram (kg), gram (g)

Time: second, minute, hour, day, week, month, year

Temperature: degree Celsius (°C)

Area: square meter (m²), square centimeter (cm²)

Volume: cubic meter (m³), cubic centimeter (cm³). Note that common practice is to use the liter, defined as 1000 cm³, for liquid measure.

Angles: degree (°). This is an exception—the SI unit is the radian, but it is anticipated that degree measures will generally be used.

Objective 5:
Teaching the International System of Units (SI)

The customary U.S. units, which are actually a collection of standard measures, are sometimes referred to as a system, but SI measures are unique in that they comprise a true system of measurement. The SI units are interrelated, simple, and easily communicated.

There are many fine sources for the development of SI in the classroom (see the April 1973 [pp. 245–88] and May 1973 [pp. 390–402] issues of the *Arithmetic Teacher*). Some principles and ideas for the effective teaching of SI units as a system follow:

1. Teach SI as a language. If children know the customary measures, they must be taught the SI measures as another language. Teach the children to "think metric" by having them measure without "translation" (conversion).

2. Develop the interrelationships between basic measures and the derived units (area, volume, velocity, acceleration, etc). Include the use of common prefixes to indicate multiples and submultiples of the basic unit.

3. Relate the teaching of SI to other ideas in which the decimal base is used, notably money and the numeration system.

4. Use charts and visual aids to show the relationships of the different SI measures.

Objective 6: Appreciation of Measurement

Appreciating the impact of measurement on our daily lives is a continuing objective of instruction. Adults may be encouraged by the fact that most of the scientific breakthroughs were made because of corresponding achievements in the art or accuracy of measurement. The child, however, may be unmoved by the mere stating of this fact.

Regular exposure to measuring experiences, approval by the teacher for completing them, and continual contact with the part measurement plays in their lives will be most effective in developing an appreciation for measurement in children. Relating the activities of the day to problems of measurement can be used to develop this objective. The space exploration program, for example, provides a wealth of material to share with children. The problems of getting to various planets involve the measurement of distance, time, angles, velocity, heat, and other things. Even the measurement of food for a space mission interests some children. Measurements used in developing high-performance engines and building bridges may stimulate the child's interest in measurement. The child may investigate the daily newspaper with a critical eye for measures and the act of measuring. Cooking, sewing, and making models interest most children and need measuring activities to make them successful. Some suggestions for activities to develop a sense of appreciation for measurement follow.

1. Ask children to bring magazine pictures that illustrate measuring devices for display on a bulletin board. If a parent is an engineer or doctor, the child may bring in pictures of unfamiliar instruments. These provide an excellent opportunity for both the children *and the teacher* to do some cooperative research. What do the instruments measure? What is the principal use for each? What dials, scales, and meters must be read to use them? Other quests will develop as answers to these questions are sought.

2. Invite parents who have some contact with measuring devices to speak to the class. Do not ignore the simple measuring tasks needed in carpentry and construction work, which also contribute to our appreciation of the measurement in our lives.

3. Establish one section of the bulletin board for a "Measurement of the Week" (or month). During the week, have each child seek illustrations of how that measure is used. Encourage them to ask parents, read the newspapers, or do research in the library—the resulting lists of applications for given measures may even surprise the teacher!

4. Assemble a slide show noting as many applications of measurement as possible, for example, parking meters, large bridges, construction

in progress, football fields, nurse's scales, and so on. What you say about the pictures will give a personal touch to things some children never consider. Good, open-ended questions can stimulate curiosity and discussion.

Objective 7: Satisfaction in Measuring

An appreciation for measurement in the environment is probably related to the satisfaction of being able to measure. It is difficult to devise specific activities that contribute to the child's feeling of satisfaction and accomplishment. Initial success, however, along with ample positive reinforcement, is the one factor that will contribute most to this satisfaction. Ask young children for simple measures. Take time to see that the children are able to use a given instrument. It is not necessary to do their measurement for them, but what they are doing does need to be appreciated and supported. Practice makes perfect—it also makes one confident in one's ability to do well. Give the children many opportunities to measure. Do not limit the measuring activity to individuals who do well. Expect *all* the children to participate in measuring, recording, and talking about what is recorded.

Give positive reinforcement to any honest attempt to make a measurement. If a child's measurement is grossly incorrect, help the child refine the techniques. Accept what measures are given, and use the divergence in measures from time to time and from child to child to illustrate the difficulty of obtaining accurate measurement.

Make the measurement lessons "hassle-free." Avoid pressure. Remember that we are interested, not in developing children's ability to recite memorized data, but in teaching them how to measure for themselves.

Ask children to measure things of interest *to them!* If measuring the desk top doesn't interest them, find something that does. Measuring for a good reason makes sense—for example, if the children are making gifts for their mothers, select projects that involve measurement. Cooking projects are fun too!

Objective 8: Accuracy and Effectiveness

Children should learn that the accuracy of their measurement depends on their care as measurers. They also must learn that the accuracy is affected by the measuring instrument. Children need to sense that even precise measurement with precision tools is not error-free. Activities to contribute to the goal of teaching accuracy and effectiveness of measurement follow:

1. Obtain several stopwatches and carefully instruct the children on their use and care. It may also be necessary to teach the children how to read the watches. Measure a course fifty meters long on the playground. Ask

a child to run the course. Appoint someone as starter and have the other children in the group time the run. Mark the finish line so it is clearly visible to all the timers. When the child has run the course, record each of the times taken by the timers. If there is some variation, discuss what the children think are the reasons for the variation. Repeat the procedure using a different child to run the course. Help the group develop a method to minimize any differences they find and to handle those differences.

2. Ask several children to measure the length of the sidewalk or fence around the school yard using metersticks, trundle wheels, or tape measures. Then ask them to compare their results and decide what to do with any differences. Which measure should be used? Should the measures be averaged? Suggest that they each remeasure the length and then compare their first reading with their second. Discuss any differences.

Both these activities may be done at the same time, using two groups of children. The teacher may wish to bring the whole class together to discuss what each group found. What were the reasons for any differences in the measurements? The teacher may then act as group recorder and list the conclusions the class makes, such as, "We must be very careful in our measuring" or "Not all of us get the same results when we measure."

Objective 9: Reading Scales and Instruments

Children should be introduced to the wide variety of common measuring instruments. Acquainting children with water meters, electric meters, gas meters, odometers, speedometers, micrometers, oil pressure gauges, and workshop gauges helps them become familiar with measurement in their lives. Clocks and thermometers come in a variety of types. Different clock-faces, including digital types, pose problems for children and illustrate the amount of instrument reading done in daily life. The parking meter and gasoline pump are as important as the measuring cup. Some activities to contribute to the goal of reading scales and instruments follow.

1. Ask local utility companies for instructions in reading their meters; some companies print instructions, which can be used in the class. Assign children to go home and read their own meters, reporting back the next day.

2. Assign the following homework assignment to be done during the week. (Be sure to remind the children each day of the assignment and ask them if they are having difficulty finding the information. A note to the parents will help the children do the assignment and involve the family as well.) "Go with your parents the next time they purchase gasoline for the car. Read the gas-pump meter and the odometer on your car and record the information, along with the date and the cost of the gasoline.

Keep a day-by-day record of the temperature outside your home. If you do not have an outdoor thermometer, you may take the temperatures given by the weather bureau. [Emphasize that you want the child actually to measure the temperature outdoors if possible.] If you have a thermostat in your house, record its temperature. Ask your parents to show you where the water meter is located and record the reading each day of the week." When this assignment is completed, have the children compare the data or record the information for the class as a whole. This activity may provide the opportunity for some computation involving differences, change, and averages. The assignment could serve as the basis for many meaningful problem-solving situations.

3. Keep a daily record of any data measured or recorded from other sources and those available in the classroom that are of interest to the children. Weather data, outside temperature, and measurements from the various meters located in the school (gas, electric, water, steam gauges in the boiler room, etc.) may be reported. Keep complete statistics on all plants and animals each day on a measurement board.

4. Most instruments require the interpreting of some form of the number line. Develop for the class activities that require work with the number line as a measuring device. The meterstick or straight rule is the simplest type. It may be necessary to demonstrate and illustrate how one estimates when a measure lies between two marked places on the rule. The idea of a number line bent around in a circle is necessary when reading circular scales. The additional idea of clockwise and counterclockwise motion must be taught with the circular scale. Learning to tell time involves all the skills mentioned. Relating the reading of circular gauges and meters to the telling of time is a good technique. Ample opportunity to use the skill should accompany instruction in how to read a number line. The more children measure with easily obtained scales, the more likely they will be able to measure with more sophisticated scales.

Objective 10: Measurement as Approximation

Measurement as approximation is often confused with accuracy in measuring. Actually, the idea of approximation is part of the physical nature of the measurement of any continuous quantity or attribute. (For a discussion of the *mathematical* nature of measurement, see the essay by Osborne.) The only physical measurement that is exact is that which deals with discrete or ordered objects. I can count the number of beans in a pile. My measurement of them is exact. For a more complete discussion of the nature of measurement, which also includes the idea of measurement as approximation, see Buckingham (1953, pp. 456–65).

Discrete or ordered measures have not been considered in this essay. Our activities in measurement have dealt with quantities whose measures are approximate. The idea of measurement as approximation is difficult to teach. Most people think we can be absolutely accurate in our physical measurements. Destroying this misconception takes many examples of measurement in which we recognize that we have gotten as accurate a measure as possible but that an exact measurement is physically impossible.

The most common measures require some estimation. For example, measure the length of a desk with a meterstick marked only in decimeters. If the length of the desk falls between two of the decimeter marks, an estimate must be made. If the meterstick is subdivided into centimeters, we shall still be faced with the problem when the measure falls between the marks. Moreover, the exact edge of the desk is as suspect as our estimate, which introduces another opportunity for error.

Work with upper-grade children involves reading scales and estimating when a measure falls between marks. Constant repetition of the process of measuring and discussion with the children help to establish the approximate nature of our measures.

Objective 11: Construction with Measures

Measuring what someone else has constructed is one thing, but constructing something that is dependent on measures is another. Reading measures from plans or recipes *and then making whatever is called for* is an important skill to learn. Some projects and activities for implementing this objective follow.

1. Construct scale models of simple or complex structures.

2. Measure some object, such as a geoboard, and construct a cardboard model.

3. Develop a "three view" picture of some simple geometric object using the measurement of the edges.

4. Construct maps of the immediate area or the school grounds.

5. Make a pitcher of water have the same temperature as the outside air.

6. Construct miniature wheels to scale using a compass and measuring stick.

7. Build a shelf or container to specifications.

8. Use a simple blueprint, available from the high school shop teacher, to construct a cardboard model.

9. Cook something, or use recipes for noncooked foods. Look for recipes whose measures are SI units. If unavailable, convert the customary ones to SI units before giving them to the children.

10. Make paper-doll dresses using SI units.

Objective 12: Continuity

It is impossible to list specific activities to develop continuity with measuring experiences. Achieving this objective depends on the teacher's commitment to the importance of measurement for the child. Some techniques and ideas to help give continuity to the measurement program follow.

1. Use incidental means to teach measurement, such as the calendar, clock, thermometer, and other devices. Keep records each day throughout the year.

2. Intersperse measurement lessons with those of number and geometry.

3. Don't wait for the textbook to dictate when it is time for a measurement lesson. Do your own—take a "measurement break" from the routines of daily number work.

4. Assign problems that include some active measurement.

5. From time to time do projects that involve measurement. Do not neglect to associate measurement with science and other areas as well as with mathematics.

6. Teach children to "think measurement" and to be measure conscious through consistent and persistent discussion and activity with measuring tasks.

7. Use measurement yourself. Learn to estimate measures. Look for examples in the newspaper and magazines. Collect pictures that illustrate aspects of measurement and take slides or pictures for use in class. Take an interest in science and other areas where measurement plays an important part. Keep measurement before the children, but above all, *think measurement!*

REFERENCES

Buckingham, Burdette R. *Elementary Arithmetic: Its Meaning and Practice.* Boston: Ginn & Co., 1953.

Copeland, Richard W. (a). *Diagnostic and Learning Activities in Mathematics for Children.* New York: Macmillan Co., 1974.

———— (b). *How Children Learn Mathematics: Teaching Implications of Piaget's Research.* 2d ed. New York: Macmillan Co., 1974.

King, Irv, and Nancy Whitman. "Going Metric in Hawaii." *Arithmetic Teacher* 20 (April 1973):258–60.

Spencer, Peter, and Marguerite Brydegaard. *Building Mathematical Competence in the Elementary School.* 2d ed. New York: Holt, Rinehart & Winston, 1966.

5

Estimation as Part of Learning to Measure

George W. Bright

Measurement can be taught more successfully if estimating is one kind of instructional activity. Since activities involving estimation not only fit into many different classroom situations but also encompass a wide variety of behaviors, frequent use of estimating is both possible and desirable. As they gain experience, students are likely to estimate more accurately and are therefore likely to enjoy estimation activities.

Probably the most common use of estimation is illustrated by this type of exercise: *Estimate the width of this page to the nearest centimeter.* This activity focuses on an identified attribute of a specified object, which is in full view, in order to obtain an estimate in terms of a known unit, here communicated by the name of the unit, *centimeter*. Since attention is called more to the measure (number) than to the unit of measure, other kinds of activities should be used to balance this one-sided approach.

Throughout this essay, attention will be given only to estimating measurements that are made up of both a number and a unit. In particular, the estimation of number or numerousness is intentionally ignored. This kind of estimation is certainly a part of measurement and is important in mathe-

The author would like to thank John G. Harvey of the University of Wisconsin—Madison for his thoughtful comments, criticisms, and suggestions.

matics instruction, but a body of literature already exists on this subject (see Bakst [1937] and Payne and Seber [1959]). The purposes of this essay are—

1. to examine the relationship between estimation and measurement;
2. to provide examples of ways of teaching estimation;
3. to show how estimation activities can be sequenced to become a useful part of mathematics instruction at all levels.

Foundations of Estimation

Estimation of measurements is the use of units of measure in a strictly mental way, without the aid of measurement tools. Estimation provides a means of applying measurement directly to the real world. Skill in estimating can improve verbal communication, for when several units of measure are used to express an idea, mental comparisons of the relative sizes of these units must be made. But before amplification for these statements is provided, several terms need to be defined.

Definitions

A *measurable attribute* (hereafter simply *attribute*) of an object is a characteristic that can be quantified by comparing it to some standard unit. For example, the mass of a rock is a measurable attribute, but the hardness of that rock is not, since the standard hardness scale is not based on a single unit. Typical attributes are length, area, volume, mass, temperature, time, and electric current.

Measuring is the process of comparing an attribute of a physical object to some unit selected to quantify that attribute. The comparison may be direct, such as when a measuring tape is held around one's waist; or it may be indirect, such as when the width of a river is measured through the use of similar triangles. Measuring also takes place when the volume of a box is computed after its length, width, and height are measured or when the circumference of a ball is measured by wrapping a string around the ball and then laying the string next to a ruler.

A *measurement* or a *measure* is the result of measuring. Both are descriptions of an attribute of an object, but there is one major difference between them. A measurement is made up of both a number and a unit of measure; for example, the circumference of the teacher's head is 59 cm. A measure is a number alone and can be derived only when the unit of measure is understood or is available for examination; for instance, in terms of the width of the ring finger, the width of the hand across the knuckles is about 4.

Estimating is the process of arriving at a measurement or measure without the aid of measuring tools. It is a mental process, though there are often visual or manipulative aspects to it. It requires that several ideas be firmly in mind: (1) the unit of measure to be used, (2) the size of that unit relative to familiar objects or to other units of measure for the same attribute, (3) other measurements in that unit, and (4) a commitment to perform the estimating so that the product is as close to the actual measurement as possible. Estimating is guessing, but the guessing must be educated. Wild guessing is not true estimating, except insofar as such behavior represents the first stages of the development of estimating skill.

An *estimate* is the product of estimating. It is the description of an attribute in terms of either a number and a unit or a number alone.

Two distinctions among terms must be clearly delineated. First, measuring and estimating are processes, whereas measurements, measures, and estimates are products of these processes. Second, physical tools are required for measuring but not for estimating.

Measurement and estimation

The mathematical view of measurement is somewhat different from the view of the preceding section. In the mathematical sense, measurement is a correspondence between an attribute of an object and a positive real number. The correspondence or assignment represents a comparison between some unit and the object in question. That is, measurement for a specified attribute is a function from a set of objects to the set of real numbers. The determination of the specific function for a given characteristic depends on the attribute and the unit selected to measure that attribute, although a simple change of variables will translate the values for a function defined for one unit to the values defined for any comparable unit. Unfortunately, the pedagogical consequences of such translations are not nearly as trivial as the mathematics of the situation might suggest.

The function is clearly many to one, for many objects can and do have the same measure, at least within the tolerances inherent in physical measurement. Mathematically, this means that it is not possible to reverse the assignment of numbers to objects. In the context of the real world, however, the assignment can, in a real sense, be viewed in the reverse way. That is, a measurement picks out the set of objects that would have that measurement. For example, in the typical measurement activity

measure the length of the table to the nearest centimeter,

the stress is on the mathematical assignment of a measure to an object. If a standard unit like the centimeter is to have meaning for students, however, there ought to be opportunities to use the assignment in the reverse direction. The following exercise might accomplish this goal:

> Use your measuring tape as a searching device to find in the room
> something that is 45 cm long.

The student is asked to locate one object that is assigned the value 45 by
the measure function for length based on the standard unit of 1 cm. This
activity helps to establish a mental picture of the size of a centimeter, or in
this example, the size of 45 centimeters. The better the student's mental
picture of 45 cm, the more efficient will be the search. At least part of the
reward for a better mental picture is a more rapid completion of the exercise. The example incorporates a safety valve in the form of the instruction
"use your measuring tape," for if students cannot visualize a length of 45
cm, the measuring tape provides a physical model that can be compared to
objects in the room.

Kinds of estimation

There are eight basic kinds of estimation. Each of the paths in figure 5.1
represents one of these. It is assumed that the unit has been selected for
each kind.

Although further subdivisions could be made in some of the categories,
the scheme of figure 5.1 is fine enough for purposes of comparison. Four
kinds of estimation, those in class A, can be used to illustrate and emphasize the mathematical properties of a measure function; four others, those
in class B, can be used to illustrate the reverse relationship between measurements and the objects that would be assigned those measurements.

In making the estimates in class A, students guess the measure for a
named attribute of an object; the object may be present or absent. Such
activities provide practice in assigning a measure to an object. Presumably,

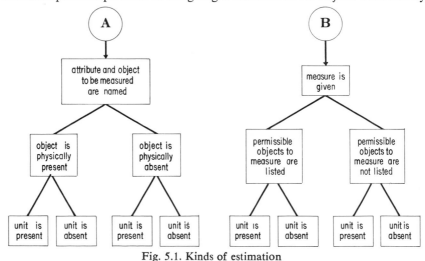

Fig. 5.1. Kinds of estimation

students mentally compare the given unit with the named object and determine the appropriate measure. With some kinds of units, however, students may not compare in the expected way. For example, in estimating the area of a rectangular object, some students will estimate the respective linear dimensions and compute an estimate of the area by applying the standard area formula. Also, when the named object is present, its position relative to the observer affects the accuracy of the estimate, at least for attributes such as length and area. Accuracy will be greater when the object is viewed straight on than when it is viewed at an angle, either horizontally or vertically.

Examples of the four kinds of estimation in class A can easily be created by satisfying each of the three conditions for each path in figure 5.1:

1. Estimate in square centimeters the area of the polygonal region shown here. [Both the object and the unit are present.]

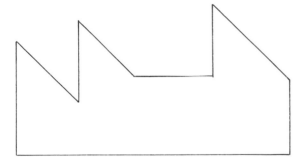

square
centimeter

2. Draw the diagonal of this page. Estimate to the nearest centimeter its length. [The object is present but the unit is not.]

3. Get a meterstick. Estimate to the nearest meter the amount of string that would be needed to make a tennis net. [The unit is present but the object is not.]

4. Estimate the volume of a garbage can in liters. [Both the object and the unit are absent.]

Students should understand that in each example they are to estimate to the nearest unit, just as they would measure to the nearest unit to check their estimates. The fourth example might be solved by mentally picturing a liter container and then mentally stacking such containers in the garbage can. Alternatively, if the garbage can is round, the height and the radius of the base could be estimated in decimeters and a volume estimate could be computed by the formula $V = \pi r^2 h$. Since the computed volume would be in cubic decimeters, the estimate in liters would have the same measure.

Since students are asked to estimate, the usual right/wrong dichotomy does not apply to their answers. Rather, there is a degree of correctness, which depends on the discrepancy between the estimate and the subsequent measurement of the object. Improving estimates should be the primary goal of these activities.

In making the estimates in class B, students either choose an object from a list of permissible objects or name an object of their choice to which a specified measure could be assigned. Such activities help students picture mentally the size of a given unit of measure. Presumably they picture the measurement in terms of the repetitions of the unit and compare this mental picture with objects (specified or unspecified) around them. Alternatively, the student may imagine a succession of objects, assign a measure to each, and compare these numbers to the specified measure. Performing the activity the first way rather than the second will probably be more useful in developing sound concepts. These kinds of estimation do not reinforce the exact mathematical concepts of measurement, but they do make units of measure more accessible to students for use in practical situations.

Examples of class B estimation may help to clarify the discussion:

1. Get a kilogram mass. Which of the following has a mass closest to 2 kg?

    football, baseball bat, hockey puck

    [Permissible objects to measure are listed and the unit is present.]

2. Which of the following would normally have a temperature closest to 5 °C?

    ice cube, candle flame, Gulf of Mexico

    [Permissible objects are listed but the unit is absent.]

3. Make a model of one square meter. Name something having an area of 6 m². [Permissible objects are not listed but the unit is present.]

4. Name something 7 dm long. [Permissible objects are not listed and the unit is absent.]

The objects that are named for the last two exercises should be chosen so that the measure of each is as close as possible to the specified measure.

One of the three conditions determining each kind of estimation is the presence or absence of the unit of measure. When the unit is absent, students must mentally picture the unit before they can begin the process of making an estimate. Consequently, students can be expected to make less accurate estimates when the unit is absent simply because an additional mental process is involved. An error in the mental picture of a unit is likely to be reflected in the estimates made for that unit.

Estimation made with the unit present can help solidify a student's mental image of the size of that unit if such activities are interspersed with those involving estimation in the absence of the unit. Once developed, the mental picture will make students more comfortable in situations requiring measuring.

Preparation for Teaching Estimation

Purposes of estimation activities

The primary purposes of including estimation exercises in the mathematics curriculum are, first, to help students develop a mental frame of reference for the sizes of units of measure relative to each other and to real objects, and, second, to provide students with activities that will concretely illustrate basic properties of measurement. Developing a frame of reference may be the more crucial for most students, since for many people the importance of mathematics lies in its usefulness in the real world. Estimation would seem to be more useful than understanding fully the mathematical structure of measurement.

Not every problem that could be solved by estimation is in fact solved that way. Some such problems are "solved" by a habitual response to a recurring situation. For example, when a new kettle is needed, the choice must be made from among a few standard sizes. It seems doubtful that the frequent cook really pays much attention to the actual capacity of the pot. Rather, the desired size is sensed intuitively. The goal of estimation in the public school is not to create such habituated responses, but to help students —who are novices in measurement—to develop skill that will give them flexibility in dealing with a wide variety of situations.

Good estimation skills come from exposure to all eight kinds of estimation previously outlined. If a shirt pattern calls for 1 m of material 90 cm wide, one must make an on-the-spot judgment of how much material 120 cm wide to buy. This is a more sophisticated problem than merely adjusting the dimensions so that the total area is the same, for the determining criterion of a successful decision is whether the pieces of the pattern will fit onto the cloth. Successful mental juggling of this sort can be enhanced by previous practice in estimating.

Developing estimation skills

Developing skill in estimating is probably best done by having students first make an estimate and then make the measurement as a check. There are no data to indicate whether students should be introduced to a unit of measure through a guess/check procedure or should be provided with opportunities to measure with the given unit before making estimates. The latter alternative seems logically more defensible, for it seems better to give students an opportunity to develop a feel for the size of the unit before expecting them to use that unit to make estimates of the same attribute of other objects. In any event, since what is being developed is the students' skills in making good estimates, it seems important that opportunities be provided in situations that are not personally threatening. Students should be encouraged to make good estimates, but they should not be penalized even for wildly inaccurate ones. Activities should be sequenced so that all students can improve the accuracy of their estimates and can develop skills that are as good as their needs demand.

Self-checking of estimates helps students develop self-correcting skills to take into the real world. That is, students learn to make their own checks. These checks will not always take the form of measuring, for such accuracy is not always needed. Sometimes a sufficient check is provided by comparing the named object to one that has already been measured.

Self-checking also helps to distinguish the act of measuring from the abstract concept of measurement. An understanding of the abstract properties of measurement should be accompanied by an understanding of the kinds of, and reasons for, errors that appear in real-world measurement data. Self-checking provides an experiential background from which errors in measuring can be explained and accurate measuring skills can be isolated, studied, and improved. Estimation exercises help students develop an appreciation for the realistic limits of the accuracy of physical measurements.

Evaluating estimation skills

Developing skill in estimating probably cannot be put in proper perspective unless some longitudinal records are kept relating students' estimates and corresponding measurements. Such records are useful, however, only if they can be interpreted in ways that help students improve their estimates. One simple procedure is to keep for each student a list of pluses and minuses—a "+" to indicate that the estimate is larger than the corresponding measurement and a "—" to indicate that the estimate is smaller. Those students who consistently have pluses (minuses) know, or at least can be told, that they need to make an adjustment in their estimates by making them smaller (larger). Unfortunately, this kind of record gives no hint as to the amount of adjustment that is needed.

An alternative procedure is to graph the data of each student's estimates on perpendicular axes, with the measurements graphed on the horizontal axis and the estimates graphed on the vertical axis. The graphed points can be compared with plotted points on the line of the equation $y = x$. Suppose, for example, that the following estimates and measurements were made in centimeters:

Estimate	7	30	40	5	15	90	110	75	60	12	20	80	25	35	2	40
Measurement	6	26	35	5	14	87	92	65	54	14	20	64	27	29	2	39

These pairs would be graphed as shown in figure 5.2. Note first that most of the points of this student's estimates lie above the line $y = x$, and second that as the measurements get larger, the amount of overestimation (not necessarily the percent of overestimation) increases. A decrease in estimates is called for, with the amount of decrease increasing as the size of the object increases.

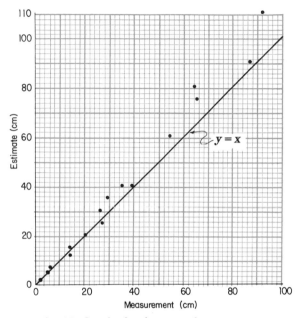

Fig. 5.2. Graph of estimates and measurements

The data could be reorganized as shown in figure 5.3 to display the percentage of discrepancy rather than the actual amount. On the average the

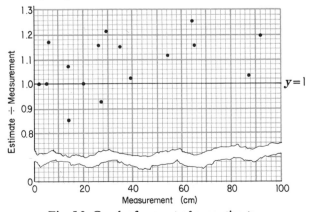

Fig. 5.3. Graph of percent of overestimate

estimates here appear to be a little more than 10 percent too large. Encouraging the student to decrease estimates by this amount for a short period of time ought to correct this bias.

Alternatively, students can be helped to remake their mental image of the unit being used. Overestimating suggests that the mental image of the unit is too small. This image might be corrected by having a physical model of the unit present while making a series of estimates or measurements. Newly measured objects could serve as referents for future estimates that would, it is hoped, be more accurate. In the remaking of the mental image, the focus of attention is on the size of the unit.

Error patterns in estimates may be related to the attribute being estimated. Patterns as clear as the one shown in figures 5.2 and 5.3 are most likely to occur for length measurement. For units of area and capacity, older students are likely to estimate linear dimensions and then compute estimates of the desired attribute. If they realize that the computed estimate can be balanced by overestimating and underestimating the different linear dimensions, then the error patterns may appear more random than systematic. Direct estimation of volume and mass is normally more difficult, since interferences arise from the shape of the container or object, the manner in which the estimate is made, and the material from which the object is made.

Estimation in the Classroom

Sample estimation activities

Mathematics textbooks for public school students include various kinds of activities designed to improve students' estimation skills or to help establish referents for specific measurements. Introductory exercises often highlight the fact that measurements are approximations; for example, a unit and a picture of an object are presented, and students are asked to fill in an accompanying blank (fig. 5.4). Such exercises also provide a good

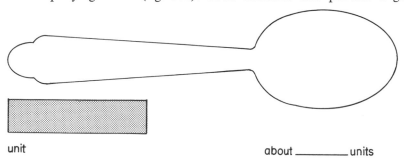

unit   about _____ units

Fig. 5.4. Sample introductory exercise

beginning for the studying of estimation, since it is the property of "about-ness" that is the focus of estimating. After the transition is made from arbitrary to standard units, the following exercises would be appropriate:

1. Without using a ruler, draw a segment that you think is 15 centimeters long. Check by measuring.

2. Guess the width of your hand to the nearest centimeter. Then find the actual measure.

Such activities help students attach meaning to names of units in the manner suggested by the Nuffield project (1967, p. 81):

> To ensure that the words [*centimeter* and *meter*] are becoming meaningful it is useful to encourage children to estimate distances before actually measuring them. The degree of accuracy of the estimates, and the use or misuse of vocabulary will help the teacher towards assessing individual progress.

Correct usage of names of units of measure tends to accompany the development of mental images of the sizes of these units. Body measurements are usually helpful in reaching this goal (fig. 5.5). Body parts can serve as referents that students can use for comparing with other objects.

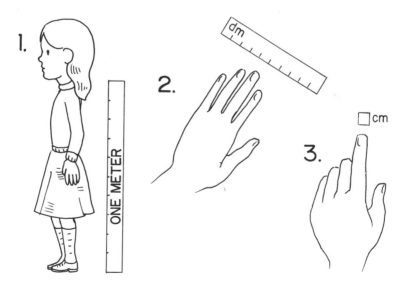

Fig. 5.5. Body referents

Although this technique is easily implemented in the classroom, it has a distinct disadvantage: as students grow, so do their body referents. Over an extended period of time, remeasuring must be done, and estimating procedures must be adjusted accordingly.

The use of referents can be externalized by having students select objects that do not change in size. The mass of a book might be used as a referent for one kilogram, and the width of a door might be taken as a model of one meter. Some objects are difficult to carry around; so students may experience some inconvenience in finding a handy referent for a specific task. Too, as students grow, their perception of the sizes of objects may change, since their size relative to the objects has in fact changed. This should not be a serious problem if estimation exercises are a regular part of the curriculum. Probably a combination of body measurement models and external models provides the most applicable set of referents. Referents should be selected not only from objects in the classroom but also from objects in the home. This will facilitate the transfer of estimation skills from the classroom to the outside world. Without such explicit facilitation, transfer may not occur at all.

Once frames of reference have been established for units of measure, the referents can be refreshed and extended through activities such as the following:

1. Complete this sentence correctly:
   A football field is about 1 _____ (meter, hectometer, kilometer) long.

2. Choose the best measurement for each object:

3 grams	25 milliliters
30 grams	250 milliliters
3 kilograms	250 liters

3. Choose the best unit for measuring each object:

kilometer	kilogram
meter	gram
millimeter	milligram

4. Which is closest to the area of a table top?
   1 cm², 1 dm², 1 m², 100 m², 10 000 m²

5. Take a large can and punch a small hole in the bottom. Guess how many milliliters of water would run out of the hole in 60 seconds. Check yourself.

Each of these problems *could* be solved by having students actually go out and measure the objects directly. Normally, however, this is not desirable, for the important objective of these activities is to develop the ability to compare objects mentally. In accomplishing this objective, teachers should not always allow students to verify their answers directly, though occasional verification can reinforce students' answers and reward their progress in improving their skills.

In spite of the variety of estimation exercises found in mathematics textbooks, two major impressions stand out in currently available series. First, estimation does not seem to be an important topic, and second, the extent and proper sequencing of activities to develop estimation skills seems to be neither understood nor appreciated. In NCTM's Twenty-fourth Yearbook, Payne and Seber suggested broad guidelines for teaching estimation (1959, p. 190):

> The practical value of being able to estimate the size of physical objects is recognized by almost all people. . . . The teacher should realize that this kind of estimation is difficult and somewhat cumbersome to teach. There are several reasons for this. It is time consuming; it may be difficult to apply a measuring instrument to the quantity; and a certain degree of mathematical mtaurity seems desirable, if not necessary. But this does not mean that we should ignore this useful aspect of estimation. It is even more reason for giving concentrated and systematic study to this topic.

In spite of this comment made more than fifteen years ago, textbooks have continued to ignore the systematic development even of minimal estimation skills. The few activities that are included in most series are scattered more or less indiscriminately throughout the volumes. Often they are segregated in brief sections headed "Explore," "Investigate," or "Extra for Fun" and thus become attachments to, but not part of, the mainstream of the content development.

Unfortunately, such superficial treatment of estimation is often accompanied by a similarly superficial treatment of measurement. This double dose of inadequate instruction leaves a quite distorted view of measurement that is usually never corrected. For example, abstraction in the process of measurement is often begun and concluded in one motion, simultaneously with the introduction of the standard units.

At the time of the initial push to rewrite mathematics textbooks to include the metric system, numerous supplementary workbooks, filmstrips, cassettes, and packages of materials were produced. The role of estimation in the instruction of these programs is noticeably different from its corresponding role in formal textbooks. Estimation is not only used more but also sequenced more carefully.

The most frequent kind of exercise is still the estimation of an attribute of a given object, such as the length of a pencil or the perimeter of a room. The reverse process is included, however; for example, students are asked to find an object that weighs the same (has the same mass) as an object of known weight (mass). The actual mass of the named object can then be estimated or measured to determine how closely it corresponds to the mass of the given object. A variation of this activity is the classifying of objects into categories such as the following: much less than one square meter, less than one square meter, exactly one square meter, more than one square meter, much more than one square meter. Estimation skills would not need to be completely developed to succeed in this activity. Nevertheless, activities of this kind would seem to be effective in developing and maintaining mental images of the sizes of many kinds of units of measure.

Careful sequencing of these activities is needed to help students build their estimation skills. Incorporating sets of sequenced activities into curriculum materials would seem to be critical in providing adequate opportunity for students to develop estimation skills, for "the need for giving pupils repeated experience cannot be overemphasized" (Payne and Seber 1959, p. 190). Estimation is not a once-in-a-while thing. It is a dynamic process that must be developed through extensive practice over a long period of time.

Games can be used both to develop and to maintain estimation skills, provided that estimation is needed to win the game. Most commercial games deal with abstractions of measurement rather than with manipulative skills. Games to teach estimation, therefore, will most likely be homemade and adapted to take advantage of the special objects that are available in a particular classroom. Team competition games and scavenger hunts are perhaps the easiest to organize. Teams can be asked either to select one object to match each of several measurements or to select several objects to match a single measurement. After estimating, students can measure the objects, and the team with the lowest cumulative error is declared the winner.

Ways of making estimates

Not all estimation skills are the same or are developed in the same way. For example, the estimation of a length could be accomplished either by imagining copies of the unit laid end to end alongside the object or by comparing the object to some other object whose measure is known. An estimate of time might be made by comparing the interval to an internalized standard or by counting the repetitions of an external event, such as the number of songs played over the radio. Estimating area can be done either directly or by first estimating linear dimensions and then computing an area estimate. The latter procedure is probably the most common method among people who are familiar with the standard area formulas, but it is doubtful that this procedure is particularly beneficial in helping people visualize the relative sizes of units of area. A more useful procedure to accomplish this end is to ask students to estimate an area by comparing it directly to one of the standard units of area. Estimating the area of an irregular shape can often be facilitated by mentally rearranging the area to form a more standard shape. The emphasis of these activities should be on visualizing the entire area at once, comparing it to the area unit most nearly the same size, and then comparing the remainder to smaller area units. The relationships among units will be stressed in this way, and more accurate estimates will probably be obtained.

A similar approach can be taken for estimating volumes of containers. Ideally, students should develop skill in estimating the capacity of an object directly, though for regular shapes (e.g., prisms, cylinders, spheres), one could estimate one or more lengths and then compute an estimate of the volume. Grocery sacks are useful in helping students visualize volumes. The shape of these sacks makes the computation of the actual volume relatively easy, and students can readily imagine filling the sacks with liter containers, since the normal purpose of sacks is to be filled. Jars (common objects that are also normally filled) have the disadvantage of often having odd shapes; so estimating volume may become confounded by a lack of skill in conserving volume. One way to provide practice in estimating volume is to fill a set of jars with different amounts of water and have students order the jars solely on the basis of the amount of water in each. This activity should of course be preceded by a lot of practice in measuring the volumes of jars. (Teachers, too, are frequently misled by the shape of a container, although this is probably not really a manifestation of the lack of conservation of volume but instead relates to their lack of practice in estimating volumes.)

The estimation of the mass of an object requires understanding the physics of the real world, so that an accurate estimate of mass can be derived. When an object is lifted, the attribute that is felt is the weight of the object, that is, the force of gravity pulling the mass of the object toward the center of the earth. Compounding the problem is that the *way* an object is lifted can alter how heavy the object feels. For example, an object lifted close to the body will "feel" lighter than the same object lifted at arm's length, and an object lifted by one hand will "feel" different from the same object lifted with both hands. Consequently, to compare objects with respect to lighter and heavier, the way each object is held should be standardized. From this type of comparison, objects whose masses have been measured can be employed as referents that can be used to generate an estimate of the mass of a given object.

Sample estimation programs

Given the variety of estimation activities and the problems inherent in developing estimation skills, how should estimation be taught? First, and most important, familiar objects need to be measured to develop a set of referents for the unit of measure being studied. This means that some experience in direct measurement should be provided immediately after the unit is introduced. Only a few measurements would seem to be called for, though, before students are asked to estimate measurements prior to actually performing the measuring. Initial estimates may be very wild indeed, but this should not discourage the use of estimation activities in the classroom.

Second, students should be encouraged to make reasonable estimates. If estimates are nothing but wild guesses, they are of no use in any practical situation.

Third, all eight kinds of estimation (see fig. 5.1) should be practiced. The net effect of incorporating all eight into a curriculum is a mutual reinforcement of skills. Although it may be possible to develop only one or just a few of these skills, skill in all eight will give students far greater flexibility in dealing with real-world problems.

Fourth, activities should provide practice in estimating both with a variety of units that measure the same attribute and with a variety of units that measure different attributes. Although there may not be much transfer of skill from one unit to another, the variety of exposure will help students develop flexibility in using estimation skills.

Fifth, estimation skill must be practiced in order to be maintained. Estimation should not be viewed as a topic that can be studied during one period of time and then never mentioned again. It should be used in a variety of situations, certainly not all of which are associated with mathematics topics. Whenever a measurement must be made—making a drawing, building something, making a map—an estimate could be made.

These five guidelines hold not only for instruction of students but also for their teachers and parents. Because of the historical lack of estimation activities in the curriculum, all three groups start with a similar degree of lack of skill. Since the groups will not progress at the same rates, the structuring and sequencing of activities should probably be somewhat different for each.

Students. The most important element of instruction in estimation for students is the provision for sufficient opportunity for the growth of their skills. Certainly as more and more units of measure are introduced, review and renewal of skills are needed. One effective way to do this is through short activities or games that provide practice in a single kind of estimation, for example, finding four objects, the sum of whose lengths is 2.5 m. Every student will not become equally proficient, but all students need to be given opportunities at each level in their own cognitive development to improve their skill. Estimation activities should parallel the measurement activities that are already part of the mathematics program, though opportunity for estimation should be provided independently if measurement is not included as part of the standard curriculum.

Estimation can be used to help teach a variety of concepts common to the mathematics curriculum of public schools. In the following examples, which are meant to be representative of the possibilities, it is assumed that each estimate is checked by measuring.

First, estimation can help solidify the concepts of number and counting as well as the concepts of more/less, larger/smaller, heavier/lighter, and so on. In making estimates of length, young students can be directed to answer the question, "How many of this unit would be as long as this object?" Answering this question requires the conceptualization of the number of units that must be put together to make the estimate. The relationship between number and unit becomes the central problem in obtaining an accurate estimate. As the size of the unit increases, the number of units needed decreases. If the concept of number is poorly understood, students cannot be expected to succeed in making such estimates.

Second, estimation can help describe the world. Amount (e.g., length, mass, temperature) is an important part of one's understanding of the world. Estimation skill can help organize this world so that patterns can be observed.

Third, computation can be practiced through estimation. Sums, products, and differences can be embedded in estimation exercises: estimate the total length of the diagonals of a rectangle; compute an estimate for the area of a rectangle; how much more does jar A hold than jar B? As decimal notation is developed to record metric measures, these computations can be performed on decimal numbers instead of whole numbers.

Teachers. Instruction for teachers can be more intense than for students. This would be reasonable even if demands of scheduling did not require it, for most teachers already possess concepts of measurement that can be used in developing estimation skills. It is probably more efficient to develop these skills through a daily workshop program of one-to-two-weeks' duration than through a weekly or monthly seminar. Extensive practice in estimating with rapid feedback provided by having the teachers measure the objects estimated seems to work well both in putting teachers at ease and in helping them develop skills. Exercises highlighting the symbolism and numerical relationships among various units can be intermingled with estimation activities.

It is important to provide instruction to teachers in a form that is readily transferrable to their own classrooms. This is perhaps easier with estimation and measurement activities than with computation activities, since many adults begin to develop their estimation skills from a conceptual base not much more secure or extensive than that of public school students. The kinds of activities useful in helping teachers develop their skills are the same kinds that are useful in helping students, though the amount of repetition required to improve skills is on the average less for teachers than for their students. All the sample exercises given in this essay are appropriate for use with teachers and can be used with many different kinds of students.

Parents. Concentration on the units of measure most frequently encountered in everyday living would seem good advice for any adult-education endeavor: meter, centimeter, kilometer, liter, kilogram, gram, degree Celsius. Estimation activities should be restricted to situations very similar to those likely to be encountered in the real world. Transfer of skill is optimized by this technique, and potential frustration is avoided. The comparison of this approach to the approach used for students should be explicitly made, so that parents understand the reasons for the differences.

Conclusion

Estimation can become a meaningful part of the public school mathematics curriculum. All that is required is for teachers to recognize its usefulness and take advantage of the variety of activities it encompasses. Students enjoy estimating; therefore, teachers need not fear using it as one of their teaching tools. By the way, how much string *does* it take to make a tennis net?

REFERENCES

Bakst, Aaron. *Approximate Computation.* Twelfth Yearbook of the National Council of Teachers of Mathematics. New York: Bureau of Publications, Teachers College, Columbia University, 1937.

Nuffield Mathematics Project. *Beginnings.* New York: John Wiley & Sons, 1967.

Payne, Joseph N., and Robert C. Seber. "Measurement and Approximation." In *The Growth of Mathematical Ideas, Grades K–12,* pp. 182–228. Twenty-fourth Yearbook of the National Council of Teachers of Mathematics. Washington, D.C.: The Council, 1959.

6

An Error Analysis Model
for Measurement

Donald R. Kerr, Jr.
Frank K. Lester, Jr.

As a topic in the school mathematics curriculum, measurement has a close and unavoidable connection with the real world—a connection that is both an asset and a liability. It is an asset in that it makes it easy to involve students in learning-by-doing situations that have natural and immediate relevance; that is, measurement lessons usually involve measuring something, and the classroom is full of interesting things to measure. It is a liability in that error is an intrinsic part of measurement in the real world; for example, in measuring the length of an object, one person estimates the length by reading a number on a ruler, but if a different person does the measuring, a different number may result. Neither of these values is exact.

Because of this inherent inexactness, different students are likely to get different "correct" answers when making the same measurement. Consequently, it is important for the teacher to understand and to anticipate sources of error in measurement so that instruction in measurement is not hampered by such seeming inconsistencies and so that the students can come to understand and control error in measurement.

Many of the ideas presented in this essay evolved in the development of the *Measurement* unit of the Mathematics-Methods Program, which was developed at the Indiana University Mathematics Education Development Center under an UPSTEP grant from the National Science Foundation.

A description or model of the measurement process called the Error Analysis Model (EAM) is presented in this essay. This model highlights the sources of error in measurement and is designed to help a teacher anticipate instructional problems in measurement and to provide a basis for teaching children about error in measurement. But first let us examine an interesting measurement activity.

Playground Measurement Activity

Suppose a sixth-grade mathematics teacher, Ms. Svrcek, challenges her students to find the area of the school playground. The teacher and students decide to form four teams of seven students each to work on this project, with each team determining its own method of solving the problem. The students are asked to keep a record of everything they do and to attempt to justify every assumption, action, and calculation they make. A diagram of the playground is provided in figure 6.1, but the students did not have access to this diagram. Notice that the boundaries of the playground very closely approximate a parallelogram.

Since Ms. Svrcek planned the lesson in advance, she spent some time

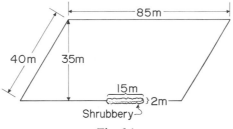

Fig. 6.1

determining the dimensions of the playground. She decided to assume that the playground had the shape of a parallelogram and determined the dimensions as shown in the diagram. Accordingly, she estimated the area to be 2975 square meters. However, since the shrubbery occupies about 30 square meters of space, she decided on an estimate of approximately 2945 square meters.

After much intense work, each of the four teams arrived at an answer:

Team 1: 3400 square meters

Team 2: Two answers—2756.25 square meters and 2727 square meters

Team 3: 3080 square meters

Team 4: 5000 square paces, or 3750 square meters

Ms. Svrcek had expected a variety of answers; so when the teams read their answers, she was not surprised. In fact, she was prepared to discuss with each team the procedure it used to solve the problems.

After discussing their procedures, she decided that team 1 had made an error in assuming that the playground was in the shape of a rectangle. They had used a tape measure to determine the dimensions of the playground and then applied the formula $A = l \times w$ to arrive at an answer. Ms. Svrcek simply asked team 1 to check the assumption that the playground was rectangular. The team then put their heads together to think of a way to find out what shape the playground had.

It seems that team 2 had correctly assumed that the playground had the shape of a parallelogram and had applied the correct formula, but they had decided to measure the width and height of the playground by pacing it off. The stride of one of the students was measured to be three-fourths of a meter. This student then paced off the playground and arrived at dimensions of 49 by 100. As a check, a different student—who also had a stride of three-fourths of a meter—was chosen to pace off the playground. However, his dimensions were 48 by 101. Consequently, they came up with two different answers: $(49 \times \frac{3}{4}) \times (100 \times \frac{3}{4})$, or 2756.25 square meters; and $(48 \times \frac{3}{4}) \times (101 \times \frac{3}{4})$, or 2727 square meters. Ms. Svrcek suspected that their results differed because the two students' strides were not uniform.

That is, it was highly unlikely that they could have maintained a stride of exactly three-fourths of a meter for every pace taken. She also helped the team recognize that no attention had been given to the space occupied by the shrubbery.

Team 3 had used a tape measure to determine the dimensions of the playground but had made an error when they measured over the top of the shrubbery, thereby adding 3 meters to their estimate of the width. Thus their computation of the area was $35 \times 88 = 3080$ square meters. This was not a serious error, but a very common type of error for sixth graders to make.

The fourth team committed a classic error. It, too, was one that Ms. Svrcek had expected. They had measured the height (50) and width (100) and had calculated the area to be 50×100 square paces, with one pace being three-fourths of a meter. Unfortunately, they made an error in calculation when they changed 5000 square paces to square meters: they simply multiplied 5000 by $\frac{3}{4}$ instead of by $(\frac{3}{4})^2$.

Ms. Svrcek's ability to anticipate the various errors that occurred suggests that she was well prepared. The Error Analysis Model presented in the next section can provide a basis for the preparation of measurement activities. The steps in the model focus attention on the sources of error in measurement, thus helping the teacher anticipate errors that occur.

The Error Analysis Model (EAM)

Looking back at Ms. Svrcek's playground activity, we see that each of the four teams of sixth graders arrived at a different number for the area of the playground. The reasons for these differences were that—

- team 1 had assumed that the playground was rectangular;
- team 2 had used students' paces as a measurement instrument;
- team 3 had laid the tape measure over the top of a bush in measuring the length of the playground;
- team 4 had made an incorrect conversion to square meters.

The errors made by the four teams correspond to four steps that occur (with varying degrees of prominence) in most measurement situations:

1. Assumptions must be made about an object in order to measure it, and any such assumption will be subject to inaccuracy.
2. Typically, a measuring instrument must be chosen, and each such instrument has built-in limitations on its precision.
3. Human judgment and skill enter into the use of any instrument. Carelessness and mistakes are involved here, but often some variability is inevitable.

4. Once numbers have been obtained, error in computation may still occur. For example, for practical reasons, it may be appropriate to round off the numbers, thus introducing imprecision.

These four steps are precisely the four steps in the EAM.

The EAM

Step 1. *Assumptions* about the object to be measured.
Step 2. *Choice* of the measuring instrument to be used.
Step 3. *Use* of the instrument.
Step 4. *Calculations,* if any, that need to be made.

A knowledge of these four steps can turn a potentially confusing situation such as Ms. Svrcek's playground activity into a rewarding learning experience. In order to explain the EAM further, let us look at another problem.

How can the volume of a rubber softball be found? One way is to immerse the ball in water and then measure the volume of water that is displaced. The assumption here is that the ball will displace a volume of water equal to its own volume. This assumption is supported by the physics of the situation, so that the step-1 error seems to be small.

Suppose that in step 2 the measuring instrument chosen is a large, transparent measuring pail that is calibrated every ten cubic centimeters (fig. 6.2). Clearly, the accuracy of these calibrations introduces some error. Moreover, since these calibrations are ten cubic centimeters apart, the instrument forces the user to make estimates that may be off by more than a whole cubic centimeter. This problem is heightened by the presence of

Fig. 6.2

surface tension on the water, which makes it difficult to determine the exact water level. All these problems can give rise to error and are introduced by the particular measuring instrument chosen; so they all contribute to step-2 error.

In the use of the instrument (step 3), error is affected by the care and skill with which the instrument is used. That is, whereas the instrument itself has certain limitations that contribute to step-2 error, the user also may introduce error—for example, by not using the instrument on a level surface; by submerging part of the hand along with the ball, thus introducing additional volume; or by misreading the calibrations on the instrument. Any such error is a step-3 error, since it is caused by the use of the instrument, not by error inherent in the instrument.

The calculations in this problem are fairly simple, but they still introduce the possibility of step-4 error. To arrive at the volume of the ball, the measurer would probably subtract the reading of the original volume of water in the measuring pail from the reading of the water level after the ball is submerged, a calculation that could be done incorrectly. Moreover, any rounding off of numbers could introduce additional step-4 error.

Thus even the simplest of measurements can involve error at one or more of the four steps in the EAM. Before getting into the classroom applications of the model, let us examine four more measurement examples.

Measurement Situations

Four measurement situations follow. Each one highlights one of the four steps in the EAM.

Situation 1

A sixth-grade class is given a glass container. Each student is asked to devise a strategy for locating on the side of the container a mark that will indicate when it is half full. The class is given a wide selection of measuring instruments, which does *not* include a measuring cup. The class comes up with the following suggestions:

1. *Use a ruler*. Measure the length of the container and then place the mark halfway up the side. (See fig. 6.3.)

Fig. 6.3

2. *Use three containers.* Fill container 1 with water. Pour approximately half the water into container 2 and mark on container 1 the level of remaining water. Pour the rest of the water from container 1 into container 3. Now pour the water from container 2 back into container 1. If the resulting water level is not at the previously estimated halfway mark, make a new estimated mark above or below the old one. Then repeat the whole process, beginning with the new halfway estimate. Keep repeating this process until satisfied.

3. *Use a formula.* Let c be the circumference of the base of the container and C be the circumference of the top. Put the mark at the level that has circumference $\frac{1}{2}(C + c)$. (See fig. 6.4.)

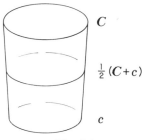

C

$\frac{1}{2}(C+c)$

c

Fig. 6.4

Of course, the sixth graders might have chosen many other strategies.

The amount of error inherent in any strategy is dependent on the validity of certain assumptions about the container; therefore, our analysis will focus on step-1 error.

1. The validity of the suggestion to measure the halfway point on the container with a ruler depends on whether or not the container has a constant cross-sectional area. If the container were a circular cylinder (vertical sides), then the half-volume line would be at the halfway point on the side of the cylinder (fig. 6.5). However, if the container were cone shaped instead of cylindrical, then the half-volume line would clearly have to be nearer the top than the bottom (fig. 6.6), and the first strategy would introduce error.

Fig. 6.5

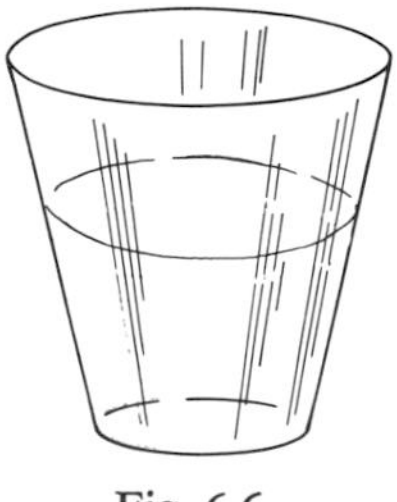

Fig. 6.6

2. Interestingly, the strategy involving the three containers is the most general one. It will work equally well for any container as long as the additional containers will each hold at least half the volume of the original one. Since this strategy does involve estimation and the repeated improvement of estimation, it does introduce some other interesting issues, but we shall not discuss them here.

3. The formula $\frac{1}{2}(C + c)$ represents a common pitfall for students. There is no particular rationale for using this fomula, except that it sounds sort of reasonable. Some gentle questions to the students on why they used this strategy would probably deal with the problem.

The merit of each of the strategies depends on the appropriateness of the assumptions about the container. Doesn't it seem reasonable that if a teacher had applied the EAM in planning this measurement activity, he or she would have stopped to ask what assumptions (step 1) the pupils might make about the container? Armed with this forethought, the teacher would be more likely to recognize the thinking behind each strategy and would be better able to analyze constructively with the class the appropriateness of each strategy. Also, if the teacher had been anxious to channel the pupils' thinking away from inappropriate assumptions, appropriate assumptions could have been suggested. It is worth noting that there are potential step-2, -3, and -4 errors for this problem, too.

Situation 2

The following situation was chosen to illustrate step-2 error and also to emphasize that the EAM has relevance outside the mathematics classroom.

Near election time a ninth-grade civics class decided to take their own poll of public opinion in the school. The object of the poll was to determine how students felt about the two presidential candidates' positions on various issues. On the issue of the economy, one student proposed that the questionnaire ask this:

Question: Which candidate's position on the economy do you prefer?

Other members of the class pointed out that this question would provide information concerning the popularity of each candidate's stand but would not provide insights into which aspects of each candidate's stand are popular and which are not. Those students suggested that the following sequence of questions be asked:

Questions: Do you agree with candidate A's claim that inflation is the main economic problem?

Do you agree with candidate B's proposal for dealing with the unemployment problem?

Which candidate's stand on international trade do you prefer?

The major differences between the different types of questions has to do with the specificity of the information that the question will yield.

In choosing any measurement instrument, one must consider the precision needed and the circumstances under which the measurement is to take place. The circumstances may sometimes make it impossible to achieve the desired precision. For example, in order to achieve the precision desired by the civics class, the questionnaire might have become longer than most respondents would tolerate. However, if the teacher and the class had been alert to the effects of the choice of the measurement instrument (step 2) on the accuracy of the measurement, the questionnaire-construction activity could clearly have been enriched.

Situation 3

No matter how well chosen a measurement instrument is, the use of the instrument (step 3) can have an impact on the accuracy of the measurement. Situation 3 illustrates this fact with a common classroom problem that might not normally be identified with measurement.

It is time to give the state mathematics-achievement test in your high school. You are one of four ninth-grade algebra teachers, and your section has been scheduled to take the achievement test during its regular class time, which is the last period of the day. Since you are particularly proud of your class, you rush in to the principal to protest that the schedule is prejudiced against your class. The principal responds that the test is standardized and therefore fair to everyone involved. You point out that although the test instrument may be a very good one, the accuracy of any measurement depends on how the measuring instrument is used as well as on the instrument itself. You argue that your class is scheduled to take the test right after physical education and right before the end of the school day and that the combination of fatigue from gym and the anticipation of the end of the school day may have a negative effect on the class's performance. You also point out that fortunately all four of the ninth-grade algebra classes have a common study hall after lunch, which would be a more equitable time for giving the test to the entire group. It is hoped that your argument wins the day and the principal becomes more sensitive to the possibility of step-3 measurement errors in future testing situations.

Situation 4

One day you give your eighth-grade class the simple assignment of finding to the nearest 0.1 cm^2 the area of a circle whose radius is 4.31 cm. Among the answers you receive are the following:

$$58.3 \ (= 3.14 \times 4.31 \times 4.31)$$
$$57.3 \ (= 3.1 \times 4.3 \times 4.3)$$
$$53.8 \ (= \ ?!)$$

These answers are different. Why? How should you deal with them?

Some preparatory thought on the teacher's part should help handle this situation. The answer 53.8 is probably the result of a calculation error. Actually, using 3.14159 for π gives rise to 58.4 as an answer; so what does one mean by "to the nearest 0.1 cm²"? It is pretty clear that most teachers could get into trouble explaining this situation to children unless considerable advance thought had been given to step-4 error.

The purpose of this section has been to help develop a better understanding of the EAM by highlighting each of the four steps in a measurement situation. A few brief comments have been made about the possible impact of the model on classroom instruction, but no general discussion has been made on how to use the model. The use of the model in preparing and implementing instruction with students in grades K–12 will be discussed in the next section.

The EAM—an Aid to "Preventive Teaching"

Planning introductory measurement activities for student application of the EAM enables the teacher to identify the points at which they may have difficulty because of imprecision, inaccuracy, or misconception. Consequently, the teacher can anticipate many of the measurement problems that might arise.

Moreover, as students gain experience with measurement they should come to know that measurements involve intrinsic as well as avoidable errors. Also, older, more experienced students should learn the various stages in the measurement process at which error creeps in. A knowledge of the EAM can help these students as they learn about the nature of measurement, and it can be a valuable aid to them as they gain facility with various measurement techniques.

Thus, the EAM has two uses in the classroom:

1. It can be an aid to teachers who subscribe to "preventive teaching."

2. It can focus students' attention on the various sources of error in the measurement process and in fact provides a means for analyzing these sources of error.

This section is concerned with the first use of the EAM—as an aid to preventive teaching. The next section will deal with ways of teaching the EAM to students.

"Preventive teaching" refers to instruction that consists of a systematic analysis of potential sources of student difficulties that could appear during instruction, a consideration of alternative ways to facilitate learning of difficult material, and the continuous diagnosis of the extent of student understanding during instruction.

Perhaps the best one-word description of preventive teaching is *anticipation*. The teacher who believes in this aspect of instruction must learn to anticipate places where students may have trouble. This is no simple task, since the best way to learn mathematics is through the students' active involvement in mathematical activities. Active involvement may be the easiest way of learning mathematics, but it increases the likelihood of floundering, misdirection, or misconception by the student. This is a particular problem with measurement, since the very nature of the measurement process makes error inevitable. Consequently, the teacher is faced with the task of encouraging student involvement in measurement activities while at the same time attempting to anticipate where students may run into difficulty. The EAM can aid the teacher in this task, since it offers a means for systematically identifying sources of error as well as areas of potential difficulty for students.

The best way to illustrate the usefulness of the EAM is to take a look at how it can be applied to teaching certain measurement topics. One of the earliest measurement topics to be dealt with in the primary grades is linear measure. Two main objectives of early experiences with linear measurement are (1) to make the children aware that instruments like rulers can be used to find out how long something is, and (2) to develop some skill in applying instruments to determine length. At the point when young children are ready to measure lengths of objects, the teacher probably wants to avoid dealing with error as much as possible. This is true because the children are not prepared to deal with the subtleties in the assumptions about the object being measured, which measuring instrument is best, or such notions as precision and accuracy. Thus, if the teacher focuses attention on the two objectives just described, an effort should be made to design measurement experiences that avoid the types of error that may interfere with attaining these objectives. The EAM can be helpful inasmuch as it equips the teacher with a way to choose activities that minimize the possibility of error. As an illustration, the teacher must try to find objects of nearly commensurable lengths, so that when children measure the lengths, they will not be confused by the fact that an object is "more than five but less than six." There is, of course, no need to choose objects that are "exactly" five or "exactly" six, but if any doubt exists as to which number corresponds to the length of an object, it is possible that the child may become confused and thereby not attain the primary objectives of the activity.

As students move into the intermediate grades, their hand/eye coordination and other psychomotor skills and their cognitive skills typically are developed well enough to warrant more concern for precision and accuracy. Also, they have had enough experience with a variety of measurement activities to be aware of the inexactness of measurement. As the students gain experience with, and understanding of, measurement, the objectives for measurement lessons also change. Instead of concentrating on making children aware that instruments can be used to measure and on developing the students' skill in using instruments, teachers should focus their concern on the various underlying assumptions involved in measurement, on developing skill in making precise and accurate measurements, and on understanding the inevitability of error in measurement. Students can learn to ask themselves such questions as, What am I assuming about this situation? Which measuring instrument is best? How precisely do I need to measure? Should I round off my answer? These questions correspond to the four steps in the EAM and point out places in the measuring process where errors may occur. Thus, as the students and the objectives change, the way in which the EAM can be used also changes. Instead of using the EAM to avoid error, teachers can use it to help students deal with error more effectively.

Before going on to teaching the EAM, we give another example of how the EAM can be used. In an article in the April 1970 issue of the *Arithmetic Teacher,* the following situation was posed (Walter 1970, p. 286):

> Recently a friend of mine had her living room floor scraped. The cost was to be reckoned by the square foot. She measured the floor to be 19½ feet long and 12½ feet wide. In order to know how much the job was going to cost, she had to know how many square feet were going to be scraped. . . . So that she wouldn't have to multiply numbers involving halves, she decided to calculate 19 × 13 instead of 19½ × 12½. . . . She realized that the calculation would have been still simpler if she had rounded off the numbers the other way; instead of changing the 19½ × 12½ to 19 × 13, she could multiply 20 × 12. . . . What a surprise she got when the two answers did not come out to be the same! . . . She was stumped. "After all," she argued, "either way, I have taken off a half foot from one side and put it on the other—it should give the same answer."

Although it would not be common to expect a student to use this procedure to find the area of a rectangle whose length is 19½ feet and width 12½ feet, it is not unreasonable to conjecture that it might happen. The possibility becomes even greater if lessons on "rounding off" have been covered recently. Teachers who employ a strategy of preventive teaching might find that the EAM tipped them off about the chance of this happening. The "friend" in the situation related earlier is guilty of a common but serious step-1 error—a wrong assumption that two shapes with equal perimeters have equal areas. Teachers who anticipate such a misconception and apply the EAM during planning could be ready with appropriate activities to clear up the misunderstanding. If the teacher did not expect this difficulty to arise, it is possible that the student's misunderstanding would not be dealt with effectively.

Teaching the EAM

One of the goals of the mathematics teacher at any level is to instill in students a view of mathematics as being an open, dynamic area of inquiry. Students should perceive that mathematical knowledge can be gained through estimating, approximating, guessing, and other nonexact methods. What better way for students to become aware of this aspect of mathematics than through the study of measurement? This would be an appropriate time for an example of a measurement situation in which estimation and approximation arise naturally and in which EAM can be a useful tool.

The question of accuracy can arise very naturally in activities involving the measurement of very thin things. (This example is taken from the very interesting and informative book *Freedom to Learn,* by Edith Biggs and James MacLean [1969].) A group of three fifth graders were asked to determine the thickness of one sheet of the paper in the storage cabinet. After mulling over the problem, the students decided to measure out one inch of paper, count the number of sheets, and then divide that number into one inch to get the thickness. The students had decided that each of them

would count the number of sheets in three different one-inch piles. They had wisely agreed on this arrangement because they saw that there was a good chance of making a counting error. The students got very tired, but through perseverance they arrived at these answers: 237, 244, and 256. They thought they had miscounted, but the task of counting the number of sheets in still another one-inch pile was too much for them. So, they decided to measure out two one-fourth-inch piles and to count the number of sheets in them. Their reason for doing this was that since the shorter piles had fewer sheets, there was less chance of making a counting error. But much to their dismay, they got two different answers: 62 and 66. They counted again and got the same answers. The problem, of course, was not due to counting but was due to the error inherent in their ruler. This led to a discussion of which was the best or most reliable result. Finally, with the teacher's help, the students concluded that only a reasonably accurate estimate of the number of sheets of paper in a pile could be made. In this example students become aware of error in measurement. The example also points to the value of equipping students with a set of guidelines to help them locate where error may occur. It is possible that if the students had been familiar with the EAM, they would have recognized the need for an instrument that was more precise than their ruler. They might also have questioned their assumption that the sheets of paper had a uniform thickness (in fact, there were three different kinds of paper—onion-skin paper, ditto paper, and construction paper).

There is no single best way to teach the EAM to students, since instructional methods are influenced by many factors, which vary from teacher to teacher and from classroom to classroom. However, the teacher who is interested in using the EAM may find these suggestions valuable:

1. Be sure the students have had considerable experience with measurement before the EAM is formally presented to them.

2. Introduce the ideas and language of the EAM informally in a variety of measurement situations (e.g., questions like "What are you assuming about this?" and "Was that the best instrument to use?" help to familiarize students with the underlying notions of the EAM).

3. Give conscious attention to making students aware of the inevitability of error in measurement.

4. Make students aware that there are several distinct steps in the measurement process and that error is possible at each step.

5. Once the steps in the EAM have been listed for the students, devise an EAM checklist for them to use as they work on measurement activities. This checklist would include the steps of the EAM.

Require the students to check off each step of the EAM as an indication that they had considered each of the steps while they worked on an activity.

6. At some point present the entire EAM formally. It does not seem likely that children will be able to infer the steps of the EAM simply by doing a variety of activities that highlight the steps of the model.

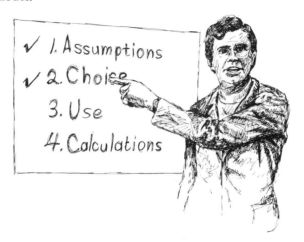

The ultimate test of the value of an instructional tool is determined by how well students learn from its use. The situation that follows is intended as a culminating activity to a measurement unit in, say, grade 7. No attempt will be made to discuss the activity in depth; instead, a few questions will be raised. Perhaps teachers can answer these and even raise some others.

Mr. Bojak paid special attention to the EAM when he planned a series of lessons on volume. A part of this unit was devoted to introducing his students to the EAM, so that they would have a means for critically analyzing their measurement procedures. On the final day of the measurement unit, Mr. Bojak challenged the class with the question, "Would a million Ping-Pong balls fit into this room?" He further challenged them to develop an argument that would convince him that their conjectures were correct.

We challenge you, the reader, to think about this problem and to come up with a convincing rationale for your own conjecture. The EAM can be very helpful in determining the reasonableness of the steps in your work. You may even want to make a checklist. For example, consider the shape of the classroom. Is it reasonable to assume the shape is a rectangular prism? Are all Ping-Pong balls the same size? How about the procedure for determining how many balls fit into a given space? (Should the whole room be filled? Of course not!) What instrument should be used? Will there be any problems in applying the instrument? How accurate will the answer be? What range of estimates is reasonable?

Experience with the EAM

Once the EAM and its uses in the classroom have become familiar, you should apply the EAM to a variety of measurement situations. Table 6.1 consists of a matrix with some blank cells to be filled in. Each horizontal row of the matrix relates to a single measurement to be made. You should fill in the potential sources of error for each of the four steps of the EAM for that measurement. (Some hints are given.) Obviously, the answers are not unique. They depend, among other things, on how one would go about the measurement. This exercise is primarily designed to stimulate thinking. The first row has been completed for you.

EAM—One of Many Models

The EAM is one of several possible models of the measuring process. Each model has characteristics that reflect its development and intended uses. The EAM was developed in an attempt to analyze the possible sources of error in basic measurement situations. Therefore, each step in the model points out an important source of error.

In Osborne's essay in this yearbook, a mathematical model of the measurement process is developed. Mathematical models are developed in order to make it possible to apply the tools of mathematics. Consequently, mathematical models employ mathematical concepts and vocabulary.

Another model describes measurement in terms that are helpful for planning instruction. Steps in this model are these:

Step 1.   Identify the attribute to be measured.

Step 2.   Select a unit quantity of the attribute.

Step 3.   Compare the entire quantity of the attribute to the unit quantity according to certain rules, arriving at a number.

One can also generate a child-development model for measurement that focuses on the manner and sequence in which children develop the mental and physical capabilities necessary for learning measurement. Such a model is particularly useful in determining the prerequisite skills and processes essential to understanding and applying measurement ideas.

It is important to recognize that each of these models is concerned with the process of measurement. Moreover, each model describes measurement in a way that serves a particular purpose. It is hoped that the EAM

TABLE 6.1

Measurement	Step 1	Step 2	Step 3	Step 4
the *growth rate* of a tree	Assume each ring represents one year of growth — but the tree may have had undiscernible rings in years of drought	Any height-measuring instrument has its limitations.	Have you ever tried to hold a tape measure up to a tree?	It's just a simple computation.
the *length* of a tangled piece of rope	Can you make an assumption that will avoid un-tangling the rope?			
the *speed* of a car				How accurate must the calculations be?
achievement in social studies	What is a good model for achievement in social studies?			
value of a used car		How accurate are the prices in the Blue Book?		
distance away of a lightning bolt			How well are the ear and hand coordinated?	
the *area* required for putting in a school baseball diamond	How far is someone likely to hit a ball?			

will help teachers plan measurement instruction so as either to avoid or to take advantage of the intrinsic discrepancies between measurements and the real objects they describe.

REFERENCES

Biggs, Edith E., and James R. McLean. *Freedom to Learn: An Active Learning Approach to Mathematics.* Don Mills, Ont.: Addison-Wesley (Canada), 1969.
Walter, Marion. "A Common Misconception about Area." *Arithmetic Teacher* 17 (April 1970):286–89.

7

SI and the Mathematics Curriculum

Daiyo Sawada
Sol E. Sigurdson

The inhabitants of North America are about to embark on an exciting adventure: the use of a metric system of measurement in all spheres of their lives. SI (Système International), the particular metric system being adopted, has the advantage of being not only a very simple but also a universal system of measurement. An estimated nine out of ten people at this time live in countries where a metric system is used. By 1985 Canada and the United States should be completely converted to using SI for measuring such quantities as length (metre), mass (kilogram), temperature (degree Celsius), light intensity (candela), substance (mole), electric current (ampere), and time (second). These are the seven basic units, but other units are derived. Time measure (seconds) and angle measure (degrees), although part of the system, are not based on powers of ten (but some scientists are advocating a 20-hour day and a 200-degree circle). North Americans will experience the adventure in two stages: (1) the period of changeover (metrication) to the new system, and (2) the period of living in a society whose every sector uses SI. It will conceivably become common to order a "five" of beer (500 millilitres), and the talk of the grocery store may develop some pseudo-SI units, say, a "newp" of butter, much as Germanic countries use a "pfund" to refer to 500 grams.

Perhaps we should be surprised that metrication (the act of changing over to using SI) will take as long as ten years. (Japan is said to have taken seventy-seven years to metricate.) Since SI is so simple, why, then, the difficulty in metrication? The answer lies in the nature of people. Old ways die hard, especially with old minds. Since the experience of many countries has been that young schoolchildren learn the system quicker than their parents, metrication in the school curriculum is fairly trivial compared to the changeover, for example, in the automobile repair shop. Educators hypothesize that schoolchildren will make the changeover to SI easily and will in fact be effective catalysts in the whole process of metrication.

Subject areas in the school curriculum such as social studies and home economics will reflect SI as it is used more in society. The mathematics curriculum, however, has the responsibility of presenting SI as a systematic scheme of measurement. SI should therefore be an integral part of the mathematics curriculum. Its introduction could simply be in terms of additional study units in, say, grades 4 and 8, or it could be introduced as a totally revised mathematics curriculum. Many believe that just as the "new mathematics" movement caused a complete revision of the curriculum, so will SI. Certainly the advocates of SI would argue that its use makes obsolete a significant portion of the mathematics curriculum. Some reflections on how SI will fit into the total school mathematics curriculum follow. In the zeal to metricate, though, the ultimate goal of a rich mathematical experience for children must not be overlooked.

The view of the current mathematics curriculum expressed herein is that traditionally an inordinate amount of time is spent on calculations; in particular, the new mathematics movement is applauded in its emphasis on basic mathematical understandings as a preparation for computation. It is also believed that the introduction of SI will further lessen the need for complicated calculations. The task in this essay, however, is not to argue for the philosophy of the mathematics curriculum but to suggest ways to maximize the benefit of SI in the curriculum. A look at characteristics of SI, characteristics of the old Imperial system, the whole area of measurement, and the general mathematics curriculum is in order so that each can be placed in its proper perspective. In order to focus on this matter, a dozen issues grouped around four basic areas of concern have been chosen:

1. The basic issue
2. Mathematical content issues
3. Instructional issues
4. Curricular issues

As each issue is discussed, it will be followed with a guideline that clearly states the authors' position.

The Basic Issue

Advocates of the metric system have long agreed on its simplicity. Many claim that as much as 25 percent of the time spent in elementary school mathematics can now be saved. Others argue that the reason British school-children fall behind other European children in elementary school mathematics can be attributed to the inordinate amount of time the British children must spend on learning a cumbersome system of measurement. Efficiency in learning will undoubtedly occur, and for a number of reasons:

1. The metric system has a smaller number of units whose names are more easily derived.
2. Calculations are much simpler because all units and subunits are related in terms of powers of ten. This is true not only for ordinary measurement but especially for calculations using large numbers expressed in powers of ten.
3. The scientist and the layman can now both use the same measurement schemes, thereby facilitating communication.
4. The metric system will be easy to learn because it, like so many other man-made systems (e.g., money and numeration), is based on the number ten.

What implications does this efficiency have for the mathematics curriculum? The most obvious implication is that measurement will be taught more easily and some complicated computations will be de-emphasized. This should allow for the introduction of new and important mathematical topics into the curriculum. Obviously, this will happen only if advantage is taken of this efficiency. A second and equally important implication is that the efficient and simple measurement system can be introduced in the early grades, allowing for a more gradual approach to measurement topics. Furthermore, it will allow some mathematical concepts and principles to develop around measurement concepts. Most of the following issues are related in some way to the efficiency that will result from the introduction of SI.

> *Guideline.* With SI, an increase in the efficient learning of all measurement concepts and measurement-related mathematical concepts can be expected. The saving in time suggests that modifications to the mathematics curriculum, even major ones, can and should be made.

Mathematical Content Issues

Domination of structures of ten

Even in premetrication days, structures of ten played a dominant role in school mathematics. In fact, because of the significance of base-ten structures, SI seemed like a "natural." However, if it is convenient for pupils to acquire competence in number work through rote learning based on the systematic base-ten aspect of our Hindu-Arabic numeration, think how convenient it would be to extend this rote learning to SI! Since it is easier to drill rather than induce students to understand, too many teachers will produce students who are able to use SI but who neither understand nor appreciate the mathematical system involved. Even if rote learning of the metric system—and the numeration system as well—can be avoided, teachers are still faced with the basic issue of the dominance of base ten in the student's thinking. No longer are there groupings of twelve for inches in a foot, of four for quarts in a gallon, or sixteen for ounces in a pound to help "liberate" students from those automatic, smooth-shifting groups of ten. Keeping in mind Dienes's principles of multiple embodiment and mathematical variability, teachers must be aware that SI produces an increasing singularity of thought through its dominating role of structures of ten. For example, the basic facts contained in addition and multiplication tables nowadays deal only with combinations up to nine, whereas combinations up to twelve (inches in a foot, a dozen) and even up to sixteen (ounces in a pound) used to be considered basic.

To help students decenter their thinking from base ten, teachers can look to new approaches in traditional school mathematics topics for some enriching mathematical concepts. Some new approaches are suggested here.

Numeration. The study of other bases was fashionable, even faddish, a few years ago. The major reason for studying other bases emerged from the idea that these experiences would give the student a better understanding of base ten. Then it became fashionable not to study other bases. Now, however, it is seen that the study of other bases has a liberating effect in helping the student overcome the domination of base ten. In approaching numeration as a "liberator," teachers would be wise to distinguish between the "fundamentals" and the "accidentals" of our numeration system. The fundamentals consist of position (or place) and rules for grouping or regrouping according to position. The accidentals consist of a specified base, specified digits for numbers up to the base, and oral names for numbers. Teachers must design activities in which the accidentals are purposefully varied while the fundamentals remain the same.

Number patterns. Searching for and exploring number patterns has become more and more a part of school mathematics. With the coming of SI, this area can be cast as a liberator by focusing on patterns not based on ten. For example, the hundred board with its 10×10 structure has been a fascinating apparatus for generating patterns based on ten, but a board that was, say, six units wide instead of ten could also be used. With such a board, some traditional mathematical processes would reflect a new format as well as reveal new patterns. For example, the sieve of Eratosthenes would look like figure 7.1. In addition to the interesting vertical and diagonal strike-off patterns, the inference that all primes are of the form

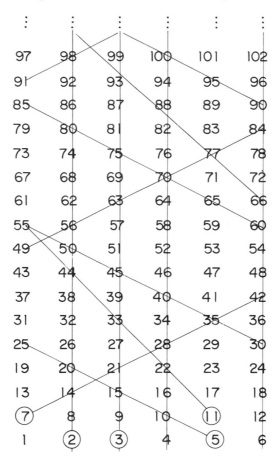

Fig. 7.1. An $n \times 6$ board used as a sieve of Eratosthenes

$6x \pm 1$ can be made (Omejc 1972). Figure 7.1 can also be used with an activity variously called Arrow Math, Lattice Maneuvers, or Mathematics on a Lattice (Page 1966). Thus on the six-column array increments are based on six rather than ten:

$$4 \uparrow = 10$$
$$4 \nearrow = 11$$
$$4 \nwarrow = 9$$
$$4 \rightarrow = 5$$

Other mathematical systems. Modular arithmetic systems are good examples of another mathematics structure that can serve as a liberator. Since clock arithmetic has been a popular modular system in school mathematics, it would be easy to introduce four-, six-, eight-, and twenty-four-hour clocks in addition to the usual mod-12 clock (see, for example, chapter 4 in NCTM's Twenty-seventh Yearbook [1963]). Another strategy for generating liberating topics consists of interpreting basic numeration concepts in more general ways. For example, the basic structure of polynomials is similar to Hindu-Arabic numeration, that is, the n-tuple $a_n a_{n-1} \ldots a_1$, consisting of the n coefficients of a polynomial $a_n x^{n-1} + a_{n-1} x^{n-2} + \ldots + a$, where $0 \le a_i < j$ for $1 \le i \le n$, can be interpreted as an n-digit whole number expressed in base j. When dealing with polynomials, teachers should point out the correspondence between whole-number numeration and polynomials, thus helping students generalize their conception of base-ten numeration while giving them a concrete interpretation of polynomials. Further application of the correspondence can then be done to advantage by relating place value to the factoring of polynomials, as suggested by Olson (1974).

Other measurement systems. The argument for examining other measurement systems is similar to that used for studying other bases. By studying measurement systems whose subunits are based on some factor(s) other than ten, the student should acquire a more comprehensive perspective on the role of ten in measurement and numeration. Of course, whereas it is suggested that time be spent on these and similar topics, it is not suggested that less time be spent on base-ten numeration and related concepts. Certainly, base ten should continue to play a dominant role, perhaps even more dominant than in the past.

> *Guideline.* **Although the dominance of the structures of ten is good for efficient thought and learning, teachers must take care to see that students, in mastering the metric system, do not become slaves of base ten.**

Decimal notation

Because SI scales are essentially isomorphic with the base-ten numeration system and Imperial measurement scales are not, some serious thought should be directed to the following question: How much, and in what way, should SI scales be allowed to determine which notational systems will be used for numbers?

For whole numbers, the picture is relatively clear. Current emphasis on various forms of expanded notation will continue. The particular form of scientific notation (e.g., $19\ 600 = 1.96 \times 10^4$) will receive added attention, and expressions such as "an order of magnitude" will be understood by elementary school children as well as by students of science. Indeed, it is hoped that the use of powers to express numerals (with or without pocket electronic calculators) will become so popular that the term *scientific notation* would become a misnomer.

For fractional numbers, the picture is much more involved. The coming of SI will mean an increase in the attention given to decimal notation and a decrease in that given to common-fraction notation. In particular, the present emphasis on common fractions (a/b) in the elementary school will be curtailed, since a primary reason for studying them has been their social utility. That utility will diminish in an SI world. However, decreased emphasis does not and should not mean the complete elimination of common-fraction notation from the mathematics curriculum. A viable position can be developed using the distinction between *algorithmic* and *conceptual* competence with common-fraction notation. Algorithmic competence means competence in carrying out the four operations in common-fraction notation. Conceptual competence means understanding the number idea in common-fraction notation. Using this distinction, the authors suggest that by the end of the elementary school years students should have both conceptual and algorithmic competence with decimal fractions but that algorithmic competence with common fractions cease to be a priority. A more modest objective for conceptual competence with common fractions seems to be in order. Thus, sixth-grade children will not be expected to divide one common fraction by another; they should, however, be able to carry out the operation using decimal fractions. Algorithmic competence with common-fraction notation would make a suitable topic for enrichment. Finally, pupils would understand the concepts of a/b in general, but computing with this notation would await a study of rational numbers in junior high school.

In summary, common fractions will be handled at a conceptional level; algorithmic competence with them will be unnecessary in elementary school. Instead, and in accordance with Dienes's principle of mathematical variability, the relationships between the notational systems (common fraction versus decimal) will be stressed. In other words, discussion will focus on the general notion of equivalent fractions and the equivalence between and within notational systems. By focusing on equivalence between systems, the teacher lessens the danger of the decimal system's becoming too dominant in the pupil's thought. Focusing on equivalence across notational systems (i.e., $1/2 \sim .5 \sim 2/4 \sim .4\overline{9} \ldots$) not only provides an excellent opportunity to study numeration as a topic in its own right but also can lead into the later study of the rational-number system.

> *Guideline.* **Common fractions should be studied only at the conceptual level; decimal numeration and its meaning should receive the major attention. Equivalence between the two systems should be established and regularly emphasized.**

Relations

Such problems in the Imperial system as the number of square yards in 6 acres or the number of pounds in 73 ounces have counterparts in SI that are indeed nonproblems. For example, 0.174 litre is simply 174 ml, or 0.730 kg is 730 g. This leads to the question, What types of mathematical processes will be neglected in SI? The answer lies in the concept of relations. The mathematical processes for handling conversion problems within the SI scales are reduced to the simple 10:1 relation, and this is handled automatically by the notation itself! This is a tremendous gain, but the atrophy of the problem-solving skills that in a metric world may find little use in measurement situations is a danger. Students will tend to handle relations such as 12:1 or 16:1 or 4:1 less frequently. Thus, the treatment of relations in general must receive some attention to avoid confining study too often to the special 10:1 relation, which works so conveniently (those automatic, smooth-shifting groups of ten again). Suggestions for giving more attention to the study of relations in general follow.

Ratio. It might be assumed that decreasing the emphasis on a/b as a common fraction calls for a corresponding decreased emphasis on a/b as a ratio. This is not true. Ratio thinking provides a primary vehicle for handling relations within the metric system, between the metric system and other systems, and between or within any system of any kind. As in other areas, care must be taken that the conversions within the metric system do not become merely rote exercises in "moving the decimal point." Although the result may be the mere movement of the decimal point, the thinking behind such a procedure must be in the ratio context—a context that serves any linear relation, metric or not. A decreased emphasis in elementary school on algorithmic competence with common-fraction notation will mean that it will no longer be advantageous (from a sequential point of view) to treat the solution of ratio equations as a natural extension of finding equivalent common fractions. Ratio ideas will stand on their own as the vehicle for handling relations, and procedures such as the "ratio test" may become more dominant than the "equivalent ratios" approach. However, new procedures may emerge based on the idea that relations, not common fractions, are the dominant topic. With the elementary school's recognition of the importance of mathematical relations, a reordering of procedures for building an understanding of ratio equations and their solution will surely be forthcoming.

Thus, the dominant embodiment of the ordered-pair idea for elementary school mathematics will be ratio rather than fraction. This is only natural, since the ratio idea expresses a relation and a relation is but a set of ordered pairs. However, the idea of fractions as ordered pairs leads to rational number (sets of equivalence classes of ordered pairs), which is a higher level of abstraction than ratio. Thus, the simpler ratio idea should be the primary ordered-pair concept for elementary school mathematics.

Indeed, the usual definition of an addition operation on ordered pairs places exclusive priority (at least in elementary school) on the definition that leads to the addition of rational numbers:

$$\frac{a}{b} + \frac{c}{d} = \frac{(ad + bc)}{bd}$$

Consequently, no priority has been given to the much simpler definition

$$\frac{a}{b} + \frac{c}{d} = \frac{(a + c)}{(b + d)},$$

which leads to the addition of vectors. In fact, it has been treated as "incorrect." This latter definition, however, is readily interpretable in the real world. For example, "a wins out of b games" in the first half of the season and "c wins out of d games" in the second half produces a total season of "$(a + c)$ wins out of $(b + d)$ games." Thus, the adding of ordered pairs leads easily to vector addition. With the stress on ratios, it may be well to give vector spaces some attention in elementary school.

Percent. With an increasing emphasis on ratio and SI, percent will play a more important role than ever, linking as it does the ratio idea with the base-ten idea. For example, 2.6 centimetres is 2.6 percent of a metre, but 2.6 inches is some "awful" percent of a foot! If percent is developed in the ratio context, it too can serve to expand the student's handling of conversion problems within the metric system.

> **Guideline.** The concept of ratio relations (as opposed to fractions) should be stressed as the important ordered-pair idea in elementary school mathematics. In part, the focus on ratios could involve informal experiences with vectors.

Rational and real numbers

In the preceding discussions, the notion of rational numbers has come up several times, and the new role of common fractions has been presented. How will this affect the study of rational numbers that has traditionally relied on the common-fraction approach? What significance does this have on the study of the reals? At the elementary school level, fractions will probably be largely a study of decimal numerals and their meaning. The concept of rational numbers as a system need not be superimposed on decimal numerals. It will be sufficient to emphasize equivalent fractions across common-fraction and decimal notation. Rational numbers would be a better topic for junior high school mathematics. Such study would be easier if common fractions were understood at a conceptual level in the elementary school. Logically, the study of rational numbers at the junior high school level could now be treated as a study in mathematical structure rather than a study of adding and multiplying fractions. With decimal numeration taking the spotlight in elementary school, the approach to real numbers in upper junior high school could be made through the non-terminating, nonrepeating decimal. The continual emphasis on the relationship between decimal and common-fraction representation begun in elementary school could culminate in the study of rational- and real-number systems in junior and senior high school.

> *Guideline.* Rational number should continue to be a major topic in junior high school mathematics, representing as it does a synthesis of the equivalence notion highlighted in the elementary school treatment of common and decimal fractions.

Estimating

Mathematics educators have long placed importance on the process of estimating. Most teachers are impressed with students who can come close to getting the answer even though they have no algorithm for providing it. With the adoption of SI, teachers can develop estimation of magnitudes by focusing on an understanding of the base-ten aspect of the measurement scales involved. Instead of making wild guesses with numbers that are difficult to manage mentally as well as to remember (e.g., 1728, 640, 36), the student will be more inclined to work with powers of ten and orders of magnitude. Estimating would become a way of improving a non-algorithmic conceptualization of quantity based on simple manipulation of a base-ten notation. The payoff for related activities in any computations would be immediate and highly desirable.

The following problem illustrates an estimation procedure:

> Find the area of a piece of cloth whose dimensions are 32 cm by 50 cm.

The estimation procedure consists of rounding off the measurements to 0.3 m and 0.5 m and multiplying to get 0.15 m^2. Notice that the estimation process involves a procedure of rounding off similar to that used in the division algorithm. The whole process depends on a firm understanding of base-ten numeration, and the process of estimating magnitudes can be built easily on base-ten relations.

> *Guideline.* Mathematics teachers should develop estimation procedures that take advantage of the structural similarity between SI and base-ten numeration.

Instructional Issues

Mathematical applications

In the past, measurement has proved a very fertile field for mathematical applications. Many an interesting day has been spent with either fence-painting or space/time problems. Although the fence-painting problem would be as interesting in any measurement system, some problems involving measurement are now rendered trivial because of the built-in structure of the metric system. For example, finding the volume of 6 pounds of water in cubic feet was a legitimate problem in the Imperial system

$$\left(\frac{1}{62.5} = \frac{v}{6}\right),$$

but finding the volume of 6 g of water in cm³ is a trivial problem in SI

$$\left(\frac{1}{1} = \frac{v}{6}\right).$$

It could be argued effectively that finding the volume of 6 pounds of water is really an unimportant problem, and now with SI, real problems can be worked on. Still, as practice for basic mathematical concepts, it is true that with SI, measurement is an area less rich for application of mathematical concepts than it was before SI, particularly for ratio thinking.

> *Guideline.* SI, as an area for applications, is beautifully simple. However, simplicity leads, in this instance, to barrenness. Thus measurement is no longer as rich an area for the application of basic mathematic skills as it once was. On the other hand, SI eliminates much of the weariness of "conversional computation," making it more feasible to tackle more meaningful problems.

A model for mathematical structure

As indicated in the preceding section, measurement has been looked on largely as an area for providing practice with basic mathematical concepts. Now, however, instead of being used as an area of application, SI can be used as a mathematical model for the development of basic mathematics concepts. In particular, students will have had prior out-of-school experiences with SI and will have some understanding of some of the basic metric scales. These scales could then be used as a model or embodiment of the base-ten numeration system or even of polynomials. In this sense, aspects of mathematics would result from a generalization of the metric scales. For example, consider a lesson where pupils needed to acquire some notion of the meaning of the location of the decimal point. An introductory activity might require pupils to cut three pieces of string into the following lengths: 1.67 m, 16.7 dm, and 167 cm. When the students notice that all three strings are approximately the same length, the equivalence of the various measures and the relationships among the subunits could be highlighted and used to give meaning to the location of the decimal point. Other concepts, such as "carrying" or "borrowing" in the addition algorithm, could also be illustrated.

Similarly, teachers should design new teaching aids that explicity take advantage of attributes measurable on a metric scale (a good example of an existing aid is the Cuisenaire rods). Attributes other than length could be focused on. Existing aids could be modified explicitly to embody the metric scales.

> **Guideline.** SI should be experienced and studied in early elementary school. This permits it to be used as a model for a variety of mathematical concepts, the most obvious of which are the various aspects of the numeration system.

Conventions

Thus far, the letter *S* in SI has been the central point of discussion. However, the letter *I* is equally important. It suggests the international and, it is hoped, the universal acceptance of the system. Universal acceptance will bring unambiguous and clearer meaning of all symbols used. It is hoped that the *I* will represent the truly international aspect of the system, so that minority groups will not "pick and choose" the system to death. Thus, if the international agreement of SI is to use the term *Celsius* to refer to the unit for temperature, then the word *Celsius* should be employed, despite the past use of the term *centigrade*.

The issue of pronunciation is one in which matters of taste begin to assert themselves. It may sound more learned and more poetic to accent the second syllable of *kilometre,* but accenting the second syllable sacrifices the meaning of the prefix for sound's sake. (Try accenting the second syllable of other subunits and see how meaning flees in the face of poetic license.)

Other points of disagreement exist on conventions of notation, but whatever the SI conventions are, we should all agree to abide by them. A major reason for adopting SI is efficiency in communication not only internationally but between the lay and the scientific communities. It will be in everyone's interest to adhere to one convention. With the introduction of SI, people are being asked to make major changes in their measurement ideas. Educators must not lead them astray by going only halfway. Certainly, some slang expressions in measurement will develop, but to keep the basic system universal in the sense of its applying to all countries and, more importantly, to all sectors of our society, is the chief goal.

> **Guideline.** The adoption of SI should be done in the spirit of *Inter*national cooperation. Educators must not split SI into an "S" and an "I" and then jeopardize the "S" by disregarding the "I."

Scope and sequence of measurement topics

SI will promote many content changes in the mathematics curriculum. In particular, new measurement topics will have to be considered. Four such topics will be SI as a total system of measurement; a comparison of SI to other systems of measurement, such as the Imperial system; the application of SI to measurement problems; and the use of measurement concepts as models to aid in understanding other mathematical ideas.

Each of these topics needs to be reviewed with regard to its sequence in the curriculum. Application of measurement as a means of applying arithmetical notions should begin very early. As arithmetic is applied through measurement, measurement models can be used to develop and illustrate arithmetical and mathematical principles. The comparison of SI to other systems of measurement can occur early and can be developed through an interdisciplinary approach. SI as a total system, however, should probably be reserved for the junior high school.

The idea of sequencing can be illustrated briefly. Suppose applications of measurement to line segments have been developed to where students measure segment AB and actually say, "The segment is 4 cm and 5 mm long."

A⊢————————————————————————B

This learning can then be applied to the problem of adding segment EF to segment CD

C   D   E————————————————————————————————F
"2 cm and 7 mm"   "6 cm and 8 mm"

to arrive at a length of either 8 cm and 15 mm *or* 9 cm and 5 mm. In this manner, SI serves as a model for illustrating both place value in numeration and the carrying principle. Later, the units of centimetre and millimetre could be compared to other "units" in other systems, such as the inch or the cubit. It might appear to students, incidentally, that the inch is a more easily understood basic unit, since it is the smallest unit in the Imperial system. Finally, when measurement is studied as a system, the common use of ten, and the common system of naming in all areas of measurement, can be discussed and related to the ease in which new units in SI are created.

> *Guideline.* **The scope and sequence of measurement topics should be based on the principle of interaction between measurement as a system of units with a concrete meaning and mathematics as a system of abstract units, each system complementing the other.**

Priorities

Obviously, time will be saved in the mathematics curriculum, since SI is easier to learn, its application is simpler, and its transfer to other mathematical topics is beneficially made. This new efficiency will leave room for more mathematics. However, some of the Imperial system should be retained both for historical reasons and for its richness in applications of ratios and proportion. Since the study of fractions will be less time consuming, the time saved could then be spent on the mathematics of the rational numbers as a number system and on ideas of vector space. Geometry can also be freed from the burden of measurement applications and be studied profitably through a less metric approach, such as motion geometry. Ratio and proportion is one area that will lose some of its most interesting applications. Here, either a more basic mathematical approach or other applications of ratio and area must be sought. An early introduction, with slow and gradual growth, of all measurement ideas, emphasizing the interrelationship between measurement and mathematics and culminating in a system of measurement, is a high priority. Through SI, measurement will become more mathematical.

> *Guideline.* **A number of areas of the curriculum other than measurement will be affected by SI. The priority should be to revise these areas to complement a metric approach. This revision will tend to de-emphasize computational algorithms and give a broader, more basic view of mathematics.**

Conclusion

The education of young people is only one area for metrication. One opinion on the metrication of the curriculum is that it should occur gradually: as society becomes more metricated, so should the curriculum. The view expressed here is that the school, especially the mathematics curriculum, can lead in the area of metrication, thereby helping to educate society. This basic premise favors a sudden, rather than a gradual, metrication of the mathematics curriculum. A second general premise is that SI is a grand invention that will make life easier but, like any invention, has implications far beyond its apparent boundaries. Some broad implications have been suggested. In particular, a dozen issues have been identified to which special attention must be paid.

Of course, teachers cannot spend their time as watchdogs of SI. School people can probably follow the system only as it is proposed. There is no doubt that the metrication movement will indeed be a blessing and will almost certainly be a boon to mathematics teaching, so long plagued with an "antimathematical" system of measurement. In a small way, the intellectual lives of schoolchildren all over North America are about to be liberated and enriched. Just as North American society is embarking on an exciting adventure, so must the mathematics curriculum be a participant in this excitement!

REFERENCES

National Council of Teachers of Mathematics. *Enrichment Mathematics for the Grades.* Twenty-seventh Yearbook. Washington, D.C.: The Council, 1963.

Olson, Alton T. "Factoring Polynomials and Place Value." *Mathematics Teacher* 67 (October 1974): 549–50.

Omejc, Eve. "A Different Approach to the Sieve of Eratosthenes." *Arithmetric Teacher* 19 (March 1972): 192–96.

Page, David A. "Maneuvres on Lattices." In *Curriculum Improvement and Innovation,* edited by W. T. Martin and D. C. Pinck, pp. 141–53. Cambridge, Mass.: Educational Services, 1966.

Standards Council of Canada. *Metric (SI) Standards Conversion Program.* Document no. CAN-P-5000B. Ottawa: The Council, 1974.

8

Parallax: A Strategy for Measurement

David E. Kullman

arallax is the apparent displacement of an object when it is viewed from two different places. A simple example of this phenomenon can be seen by holding a pencil vertically at arm's length. When looked at with one eye at a time, the pencil's apparent shift in position relative to distant objects in the background is parallax. For an example of parallax on a larger scale, sight two objects that are in a straight line with yourself, but at greatly different distances. Move a few steps at right angles to the line of sight, view the objects again, and notice that they are no longer in alignment. This apparent displacement is parallax.

The subject of parallax is seldom found in mathematics courses; yet it can be an excellent vehicle for learning and applying concepts of measurement, angle relations, similar triangles, and trigonometric ratios. It also provides an incentive for students to interact with the physical world, and it can be used as the basis for laboratory activities in mathematics. Some activities that are appropriate for mathematics classes in either junior or senior high school will be illustrated. Students who have not yet studied trigonometry can replace any required trigonometric calculations with carefully made scale drawings.

138

The Geometry of Parallax

Parallax refers to an apparent displacement of an object; more precisely, it is defined as the difference in direction of an object, as viewed from two places. This difference in direction can be expressed as an angle, and the two viewing positions can be represented as endpoints of a line segment, called the base line (fig. 8.1). It is best to choose a base line that is

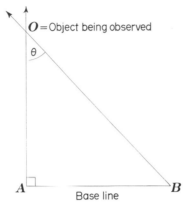

Fig. 8.1. θ = parallax

perpendicular to one of the lines of sight, so that a right triangle is formed. It should be clear that the parallax of a given object will increase as the base line is made longer, and, for a fixed base line, the parallax will decrease as the distance of the object is increased. From figure 8.1 the following relation can be easily derived:

$$\text{tangent of parallax} = \frac{\text{length of base line}}{\text{distance from base line to object}} = \frac{AB}{AO}.$$

How can the parallax of an object be measured? One obvious way is to measure the angle formed by the base line and the line of sight at point B. The parallax is the complement of this angle. This procedure, however, is not always feasible. When the base line is very long, or when obstacles are in the way, A may not be visible from B, and hence the angle at B cannot be measured accurately. There are also situations where the base line is not perpendicular to either line of sight from its endpoints, and a different strategy is required.

It is a common procedure to measure the parallax of an object O by choosing another object D at a much greater distance from the observer than O. In particular, D is chosen so far away that its parallax is too small to be detected. That is, the parallax of D is taken to be zero, and the two lines of sight, AD and BD, are considered parallel, as in figure 8.2.

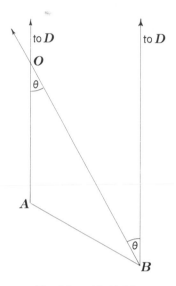

Fig. 8.2.   *AD || BD*

If the points A, O, and D are not collinear, $\angle DAO$ and $\angle DBO$ can be measured and the parallax of O expressed as their sum (fig. 8.3) or difference (fig. 8.4). This strategy does not require any of the lines of sight to be perpendicular to the base line.

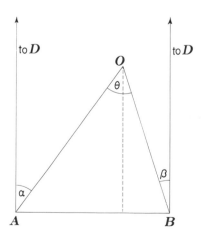

Fig. 8.3. $\theta = \alpha + \beta$

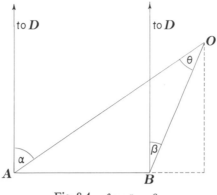

Fig. 8.4. $\theta = \alpha - \beta$

Applications

The concept of parallax has important applications in astronomy and with optical range finders. In astronomy, two types of parallax measurements are employed. The *geocentric* parallax of a body in the solar system is the difference in its direction as seen (1) from the center of the earth and (2) from an observing station on the earth's surface. Measurments are made simultaneously from two places on the earth a known distance apart. The observed parallax is then standardized by calculating the parallax that would have resulted if the base line had been the earth's equatorial radius and the body being observed had been on the horizon (fig. 8.5). The geocentric parallax of the moon has been found to be 57 minutes 2.7 seconds, or 0.951 degrees.

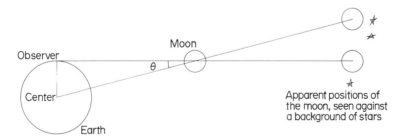

Fig. 8.5. Geocentric parallax of the moon

The *heliocentric* parallax of a star is the angle subtended at the star by the semimajor axis of the earth's orbit. This is determined by studying photographs taken six months apart and observing the apparent displacement of the star against a background of stars so remote that they do not

show any measurable parallax (fig. 8.6). Even the closest stars are so
distant that their parallax never exceeds one second of arc.

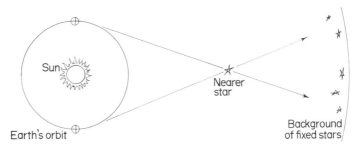

Fig. 8.6.   Heliocentric parallax of a star

Activities

Concept of parallax

Materials needed: meterstick

Have the students work together in pairs. Have one student stand five
to ten meters from the chalkboard, holding a pencil (denoted *O*) vertically
with arm outstretched. While this student sights the pencil with one eye at
a time (*A* and *B*), instruct the partner to record the apparent positions of
the pencil by making two vertical line segments (*A'* and *B'*) on the chalkboard (fig. 8.7).

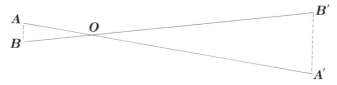

Fig. 8.7.   Eyeball-to-eyeball parallax (top view)

Let students measure the distances *AB* (eyeball-to-eyeball), *AO* (eyeball-
to-pencil), and *A'B'* (between the lines on the chalkboard). By an easy
application of similar triangles, the students can compute the distance
from the pencil to the chalkboard, *OA'*. The parallax, *m ∠ AOB,* can be
determined by means of a scale drawing.

Optical range finder

Materials needed: meterstick, protractor, two pocket mirrors (mounted on
                    blocks of wood so they will stand vertically)

Have two students work together on this experiment. Set the two
mirrors on a table at the ends of a measured base line *AB*, approximately

0.5 meter long, and a convenient distance (approximately five meters) from a chalkboard (fig. 8.8). Have one student draw a vertical line on the chalkboard at O so that AO and AB are perpendicular. (Instead of a line on the chalkboard, any other vertical reference line may be used.) Let the other student set the mirror at A at a 45° angle with segment AB. Have the first student adjust the mirror at B so that the observer at A sees the line at O (over the top of the mirror) coinciding with its image as reflected by the mirrors. Ask the students to measure the angle α, formed by AB and the plane of the mirror at B. Using the law of reflection, show that $m\angle ABO = 180° - 2\alpha$. If α is known, the parallax of O is simply the complement of α. With these data, the students can determine the distance AO either by using trigonometry or by making a scale drawing.

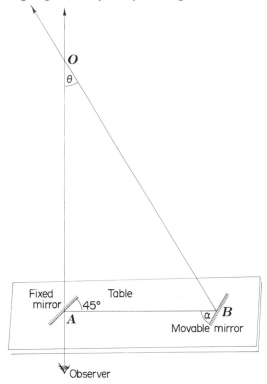

Fig. 8.8. The geometry of a range finder (top view)

Students should realize that a small error in measuring the angles at A or B can cause a significant error in determining AO. Ask them to repeat the experiment several times (possibly by a different pair of students each time) so that the mean distance AO can be obtained. Depending on the sophistication of the class, this might be a good time to discuss the nature of experimental error and how it must be taken into account whenever measurements are made.

Model T parallax viewer

Materials needed: a strip of pine about 1 cm by 5 cm by 45 cm, a strip of balsa about 1 cm by 2.5 cm by 35 cm, rectangular cardboard about 4 cm by 5 cm, a paper clip, small screws or nails, two straight pins, a meterstick

Let the students work individually or in pairs. Have them construct the parallax viewer by gluing or nailing the strip of balsa at one end of the strip of pine to form a T (figs. 8.9 and 8.10). Use screws to fasten the paper clip at the intersection of the pine and balsa strips. Fasten the cardboard rectangle at the other end of the pine strip and make a pinhole through its center. This pinhole and the paper clip will be used as "sights" in aiming the viewer (fig. 8.11).

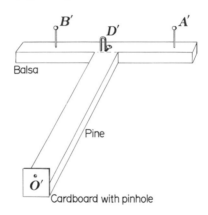

Fig. 8.9.   Details of the Model T parallax viewer

Fig. 8.10.   The Model T parallax viewer

Fig. 8.11.   Using the Model T parallax viewer

Explain that when this parallax viewer is used to determine the distance to an object O, it is necessary to choose a second object D very far away, as in figures 8.2–8.4.

Have a student stand at A, fix the object D in the sights of the viewer, and place a pin in the balsa at A' in line with the object O. Instruct the student to move an appropriate distance (to be determined experimentally) from A along a base line perpendicular to AD. Call the other end of the base line B. At B repeat the procedure of fixing D in the sights of the viewer and placing a pin in the balsa at B' in line with O. From the diagram in figure 8.12, it is easy to show that triangle AOB is similar to the triangle $A'O'B'$ formed by the two pins and the eyepiece of the viewer. It is then a straightforward calculation to determine the distance from O to the base line AB. The actual parallax (measure of angle AOB) can be determined either by making a scale drawing of triangle $A'O'B'$ or by trigonometry.

Students will enjoy using a Model T parallax viewer both indoors and outdoors. If care is taken in constructing and using it, distances up to a kilometer may be measured. But it is always important to make suitable choices for the reference point D and base line AB. Students who have

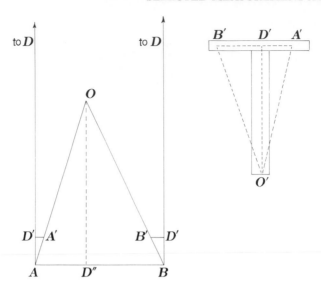

Fig. 8.12.   Using the parallax viewer: $\triangle AOB \sim \triangle A'O'B'$

an intuitive feeling for choosing these two features probably understand the principles of parallax. Also, in this activity, students and teacher should discuss errors in measurement and how to work with approximate numbers.

9

Using Photography to Measure Velocity

Paul Wm. Rahmoeller

C ommonly used methods for determining velocities of fast-moving objects usually have one of the following four disadvantages, especially in sports events: (1) The situation is artificial, (2) an unnatural force acts on the moving object, (3) the recorded velocity is actually an average velocity over a relatively long period of time, or (4) the method of measurement is quite expensive.

In this essay a relatively inexpensive, portable, and reproducible method for finding average velocities over a short period of time, one that does not interfere with the natural course of the velocities, will be presented. The exercise involves a small amount of skill in photography and can be used as a special project with mathematics or science students.

The example provided here involves finding the velocity of a softball pitched during a game. The object is to photograph the pitch at a shutter speed that will not "stop" the action of the ball—when a print is produced, the ball will appear as a blur, as in figure 9.1. (Do not be alarmed that the negative or photograph looks out of focus. Only the softball should be in focus; but it is not, because it was moving.) If the length of this blur and the shutter speed of the camera can be determined accurately, then the actual displacement of the ball during a time interval can be found. The

147

Fig. 9.1.

average velocity of the ball during this time interval is the quotient obtained by dividing the actual displacement by the time interval.

In order to find the actual displacement of the softball, an enlargement of the photograph is made to show the entire blur, as large as possible or practical (figs. 9.2 and 9.3). The height of the blur gives the diameter of the ball in the photo. The length of the blur is then found and multiplied by a factor equal to the quotient of the actual diameter divided by the height of the blur. One diameter of the softball must be subtracted from this length to account for the width of the softball, which is not part of the displacement.

Fig. 9.2. A ball traveling 26.3 km/h, or approximately 16 miles an hour, photographed with a shutter speed of 1/15 second

Fig. 9.3. A ball traveling 67.3 km/h, or approximately 42 miles an hour, photographed with a shutter speed of 1/15 second

For example, a softball 9.6 cm in diameter is photographed at a shutter speed of 1/15 second, and the blur obtained on an enlarged print is 0.70 cm high and 4.25 cm long (fig. 9.2). The actual displacement of the softball would be

$$\left(\frac{9.60}{0.70} \times 4.25\right) - 9.6 = 48.7 \text{ cm.}$$

Since the shutter speed was 1/15 second, the average velocity for that time period would have been

$$48.7 \text{ cm/} \tfrac{1}{15} \text{ s} = 730.5 \text{ cm/s} = 26.3 \text{ km/h.}$$

This is approximately 16 miles per hour (a slow pitch!).

An alternate formula for velocity in kilometers per hour is

$$\text{velocity in km/h} = \frac{(L-H) \times D}{H \times T} \times 0.036,$$

where

L = length of blur in centimeters

H = height of blur in centimeters

D = diameter of softball in centimeters

T = shutter speed in seconds.

Thus:

$$\frac{(4.25 - 0.70) \times (9.6)}{(0.70) \times (1/15)} \times 0.036 = 26.3 \text{ km/h}$$

If feasible, a meterstick or other calibrated measuring device can be used instead of the height of the blur to find the multiplication factor for these calculations. Since the measuring device would be stationary, it should be positioned near the path of the object to be photographed, and it should be included in the enlargement. If, because of lighting and background problems, a clear picture of the blur cannot be obtained, then either some other method of measurement must be used or the measurement must be made under artificial circumstances.

Some other factors must also be considered. Although most camera dealers assure us that shutter speeds are consistent and accurately marked, this must be checked. There are several expensive and elaborate ways to check a camera's shutter speed. However, one homemade method is to place a very thin strip of adhesive tape along the radius of an unwanted phonograph record and spin the record on a turntable at the fastest setting, usually 78 revolutions per minute. Without the drag from a stylus, this will not be an accurate rate of spin. The spin rate must be timed over several three- or four-minute periods to check for an accurate and consistent rate for the record.

When the rate is reduced to revolutions per second, the dimensions should be changed to degrees passed through in a second by multiplying by 360, the number of degrees in a full turn. At 65 revolutions per minute, camera shutter speeds of 1/2 second, 1/4 second, 1/15 second, and 1/30 second would produce blurs on a photo of 195.0 degrees, 97.5 degrees, 26.0 degrees, and 13.0 degrees, respectively. These data and subsequent calculations are based on figure 9.4.

Fig. 9.4. A record revolving at 65 r/min, photographed with a shutter speed of 1/15 second

To check faster shutter speeds, a variable-speed drill can be used to produce a faster rate of spin. The phonograph record can be mounted on a sanding disk and used as previously described. Once again, the spin rate of the drill and the smoothness of the drill's movement would need to be checked first. This method can also be used for checking shutter speeds on cameras lacking shutter-speed indicators. A camera without variable shutter speeds and variable lighting controls will severely limit the quality of the photos produced and the range of velocities that can be measured.

If a camera is found to have a consistent, but not precise, shutter speed for each of the settings, it will be satisfactory with this method of finding velocities. Before a moving object is photographed at a certain shutter speed, a photograph of the spinning record should be made as described. After the moving object has been photographed at the same shutter speed, a second photograph of the revolving record should be made. After prints are made, the shutter speed can be calculated from photos of the revolving record by knowing the actual rate of spin and the degrees through which any point on the record revolved while the shutter was open.

Skill in timing the precise moment to take a photo and in setting light controls will need to be developed. An observer should also try to get into a position that is perpendicular to, and distant enough from, the path of the moving object in order to catch the entire displacement. If these precautions are taken and these skills acquired, then a readable picture showing the entire blur can be obtained. (The distortion of length due to the angle at which the picture is taken will be negligible.) A dark object, however, will always be hard to photograph except under ideal conditions.

Most high schools have the appropriate photography equipment needed for this exercise. Science club members not only could calculate velocities for most of their high school sports activities but also could find the velocities for model airplanes and rockets.

Some extensions of this method for finding velocity might be used on a video tape of Hank Aaron's record-breaking home-run hit, a motion picture of a tennis ace by Rod Laver, or a stop-action picture taken during a jai-alai match. Each incident could illustrate how fast the ball traveled at a specific time.

10

Using the Carpenter's Square

Mary Kay Smith
Sidney L. Rachlin

The activities described in this essay will acquaint students who are familiar with the techniques of using compass and straightedge with an alternative approach to making constructions. These activities are supplementary; they are not intended to replace conventional approaches. The instrument used is a simple adaptation of the carpenter's square (or framing square, as it is sometimes called), a tool used to solve layout problems that might otherwise be extremely difficult.

Construction of the Carpenter's Square

In order to construct the carpenter's square, scissors and a sheet of posterboard two decimeters by three decimeters will be needed. Cut a strip of posterboard approximately four centimeters wide, as shown in figure 10.1. The longer arm will be referred to as the *blade;* the shorter arm is the *tongue*. The *heel* is the point where the outside edges of the square meet.

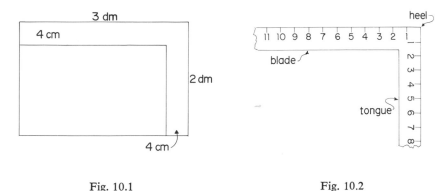

Fig. 10.1   Fig. 10.2

Begin at the heel and mark off both arms in unit lengths (e.g., centimeters, inches, paper clips, etc.). The finished square should appear as shown in figure 10.2.

Dividing a Line into a Number of Equal Parts

The standard construction technique for dividing a given line into a number of equal parts is paralleled by the carpenter's use of the square to rip a board into a number of equal strips. This approach may be adapted to the classroom by giving students a piece of cardboard of a specified width (e.g., four units) and having them use the square to divide the cardboard into a number of strips, say, five, of equal width.

For this activity, the carpenter's square and a piece of cardboard four units wide and twelve units long will be needed. Place the square across the cardboard with the heel at one edge of the cardboard. Line up the five-unit mark on the blade of the square with the other edge of the cardboard (see fig. 10.3). Place a point on the cardboard at each unit mark of the scale. Move the square to a new position on the cardboard and repeat the process (fig. 10.4). Join the corresponding points to form a

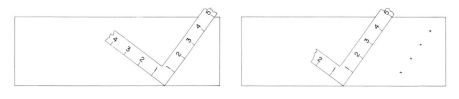

Fig. 10.3   Fig. 10.4

series of parallel lines, which will divide the cardboard into five strips of equal width (fig. 10.5).

Fig. 10.5

A possible follow-up activity is to vary the width of the cardboard and the number of desired strips. For example, give the students a piece of cardboard five units wide and ask them to use the square to divide the cardboard into two strips of equal width. In this situation one procedure would be to line up the six-unit mark on the edge of the cardboard and use the three-unit mark as the division point. Of course the eight-unit mark could be used with the four-unit mark as the division point, or the ten-unit with the five-unit mark, and so on.

Have the students discuss why the method works. If they are acquainted with other methods for dividing a line into equal parts, have them make a comparison between those methods and the use of the square just described.

Bisecting an Angle

It is assumed that students are acquainted with the method of bisecting an angle by using a compass and straightedge. These students should find it interesting to learn that an angle can also be bisected by using the carpenter's square. Again, they should discuss why this particular method works.

To bisect an angle with the carpenter's square, mark off from the vertex a convenient equal distance (e.g., 6 cm) on each ray of the angle (fig. 10.6). Place the heel of the square away from the vertex and line up the square so that the heel is the same distance from the two marks made on the rays of the angle (see fig. 10.7). Place a point at the heel of the

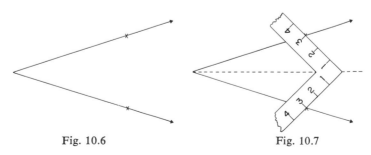

Fig. 10.6   Fig. 10.7

square, and then draw a line through the vertex of the angle and this point. This line will bisect the angle.

Other Constructions

The carpenter's square may also be used to construct the perpendicular bisector of a given line segment. For example, to construct the perpendicular bisector of line segment *AB*, first use the square to draw a perpendicular at each endpoint of the segment, as shown in figure 10.8. Then use the technique described in the first activity to divide segment *AB* into two equal parts, as shown in figure 10.9. Segment *DEF* will then be the perpendicular bisector of segment *AB* at the midpoint *F* of $\overline{AB}$ (see fig. 10.10).

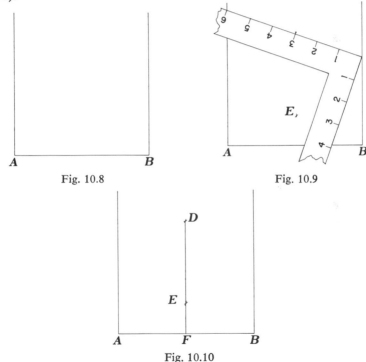

Fig. 10.8   Fig. 10.9

Fig. 10.10

This technique, in conjunction with the angle-bisecting technique, may be used to construct angle bisectors, medians, and perpendicular bisectors of the sides of a given triangle. In fact, any basic geometric construction that involves determining angle bisectors, perpendiculars, or segment bisectors can be accomplished by using the carpenter's square in lieu of the more traditional compass-and-straightedge constructions.

11

Measurement on Surfaces

John K. Beem

Whenever such surfaces as spheres, cylinders, and tori are introduced in a geometry class, the question of how to define simple geometric quantities arises. The student would like some idea of how to measure areas, angles, and distances. In this essay the problem of measuring distances on a surface will be considered, geodesics will be introduced, and several ways of finding geodesics experimentally will be given. Surface area will be discussed for a class of surfaces known as *developables*.

Assume that a surface S is given as a subset of ordinary Euclidean space. Two viewpoints may be taken, the first of which is called the *extrinsic* viewpoint. Think of three-dimensional beings looking at a two-dimensional object S that is embedded in a three-dimensional world. (The surface S is only two dimensional, since a surface has thickness zero.) The second viewpoint is the *intrinsic* viewpoint, illustrated by imagining two-dimensional beings restricted to living on the surface. For example, if they wish to go from one point on the surface to another, they must do it by traversing a path that lies on the surface rather than by traversing a path in three-dimensional space. In studying surfaces, one often shuttles between the extrinsic and the intrinsic viewpoints. Each viewpoint has something different to offer.

Geodesics

In ordinary plane Euclidean geometry the straight line segment joining two distinct points is the unique curve of shortest length from the first point to the second. The situation for surfaces other than planes can be more complicated.

Consider the earth, which is idealized as a sphere, and ask, Does there exist on its surface a unique curve of shortest length joining the North Pole to the South Pole? The answer is no. Actually, there are an infinite number of curves of shortest length joining the North Pole to the South Pole. These curves are semicircles of the great circles passing through both poles. They lie on meridians of the earth.

On a given surface, a curve that is the analog of a straight line in the plane is called a *geodesic*. For example, on a sphere the geodesics are the great circles. These are formed when a plane through the center of the sphere intersects the sphere. This same example of the sphere shows that for surfaces in general a geodesic cannot be defined simply as the curve of shortest length joining distinct points. However, a geodesic can be defined as a surface curve whose every sufficiently small piece is a shortest path. A geodesic can be viewed as a curve on the surface made up of small overlapping segments, each of which is a shortest curve on the surface. It must be pointed out that a given geodesic with endpoints p and q relatively far apart might not be the shortest curve joining these widely separated points.

Consider a right circular cylinder. The geodesics on the cylinder are the helices that wind around the cylinder, the circles that go around the cylinder, and the straight lines (generators) that are parallel to the axis of the cylinder. These curves are illustrated in figure 11.1.

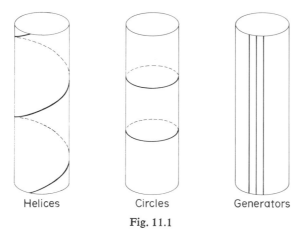

Helices   Circles   Generators

Fig. 11.1

Let p and q be two points that lie on a common generator of the cylinder. The shortest curve from p to q is a segment of the common generator. However, there are many helices that wind around the cylinder and pass through both p and q. Each of these helices has a different pitch, and each is a geodesic through p and q that is longer than the geodesic lying on the generator.

Distance

On a connected surface S, the distance between two points p and q can be defined as the infimum of the lengths of curves on S that join p and q. This definition of distance makes S a metric space.

For ordinary surfaces, the distance between two points is the length of a shortest geodesic joining the points. Consequently, for many surfaces one can first find the geodesics and then use this information to calculate the distance between the points. A geodesic of shortest length joining two points is called a *minimizing* geodesic.

Experimental Methods

There are three methods, easily demonstrated in the classroom, of experimentally finding the (approximate) geodesics on a surface. The first two methods will be called the string method and the paper-strip method. These procedures work for all ordinary surfaces. The third method, called the cut-and-bend method, works only for a special class of surfaces.

The string method consists of placing the ends of a string on a surface at points p and q and then pulling the string tight. Naturally, the string must remain in contact with the surface at all points between p and q. This method works very well on spheres and cylinders. It can be used experimentally to verify that the previously mentioned curves on these surfaces are indeed geodesics. This method is illustrated in figure 11.2.

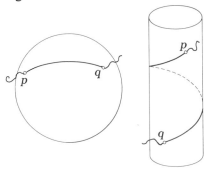

Fig. 11.2

A good classroom demonstration can be given using only a globe of the earth, a string, and a ruler. Choose two points p and q on the globe. For example, let New York be p and London, q. Pull the string tight with one end on p and the other on q. The string lies along the great circle route from p to q, the most direct route for a plane to fly. Measure the length L of the string from p to q and then measure the length Q of the string with one end at the North Pole and the other at a point on the equator. The length Q represents 10 000 000 meters. Consequently, the distance d from p to q on the real world is given in meters by

$$d = \frac{L \times 10\ 000\ 000}{Q}.$$

In this formula the lengths L and Q must be expressed in the same units.

The physical reason the string method works is that a taut string on a smooth surface tends to minimize length as the tension in the string is increased. The string must always remain in contact with the surface. This method does not work very well with surfaces that have a number of waves, since the string tends to pop off the surface when it is pulled tight.

The paper-strip method uses a long, narrow, rectangular strip of paper that has a straight line C down the center. This paper strip is placed on the surface and flattened against it as much as possible. The center line C then represents a geodesic on the surface (see fig. 11.3).

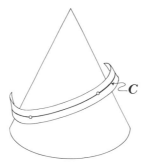

Fig. 11.3

In this method, each small piece of the paper strip is being used to approximate the tangent plane of the surface. The normal to the surface is perpendicular to the tangent plane and, hence, to the paper strip at each point of C. Furthermore, the principal normal to the curve C is perpendicular to the paper strip. The fact that C is a geodesic follows from the fact that these normals are parallel.

It is interesting to compare these two methods. The string method is an intrinsic method, since the increase in the string's tension moves it sideways along the surface until its length has been minimized. (The string method could be used by two-dimensional beings living on the surface.) The paper-strip method is an extrinsic method, since it uses normal vectors that point perpendicular to the surface and to the paper strip. The paper-strip method makes use of the fact that the surface is a subset of ordinary Euclidean space.

Developable Surfaces

An important class of surfaces, known as the developables, are surfaces that may be cut and then bent in such a way they lie flat on a plane. The bending must be done without any stretching or shrinking of the surface. Cones and cylinders are developables; spheres, tori, and paraboloids are not.

Since developables may be laid flat after a few simple cuts, they are well suited for classroom experiments. In particular, it is easy to find the surface area and the geodesics of a developable.

Consider the following experiment. Take a right circular cylinder made of paper with helices marked and cut along a generator. Lay the paper flat on a plane by bending it without stretching. The helices of the cylinder will lie on straight line segments of the plane. If the cylinder had height H and a radius R of the base, it is easy to verify that the flattened paper is a rectangle of sides H and $2\pi R$. Clearly, the lateral (side) area of the cylinder is $2\pi RH$. This experiment is illustrated in figure 11.4.

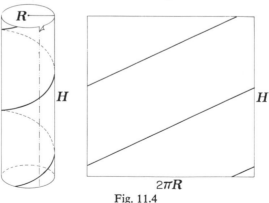

Fig. 11.4

In general, the geodesics of a developable may be found by first cutting and bending the surface until it is on a plane, then drawing straight lines on the flattened surface, and finally pasting the surface back together. An easy assignment for students is to have them find the geodesics on a given paper cone.

Another simple cut-and-bend experiment can demonstrate the formula πRL for the lateral area of a right circular cone of slant height L and radius R of the base. Cut the cone along a generator and lay the paper on a disk of radius L, as shown in figure 11.5. The circumference $2\pi R$ of the base of the cone lies on part of the circumference of the circle of radius L. The area A of the cone is then in the same ratio to the area πL^2 of the disk as the length $2\pi R$ is to the circumference $2\pi L$. This yields $A = \pi RL$.

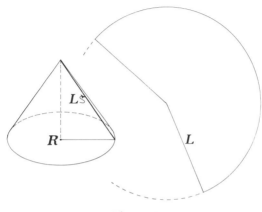

Fig. 11.5

The Möbius band, shown in figure 11.6, is one of the most interesting developables. To construct this surface, give one twist to a strip of paper and paste the ends together. This surface is "one sided." It is impossible to color one side red and one side blue. The geodesics are easily found by drawing straight lines on the strip of paper before the ends are pasted together.

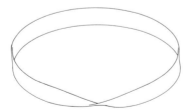

Fig. 11.6

Nondevelopable Surfaces

It is an interesting fact that a sphere is not a developable. Consequently, a sphere cannot be flattened onto a plane after a few simple cuts followed by bending without stretching or shrinking. Thus a completely accurate flat map of the earth cannot be made, for any flat map of the earth reflects some distance distortion that varies from point to point on the map.

In order to demonstrate that a sphere cannot be flattened on a plane without some stretching or bending, a hollow rubber ball may be used. When the rubber ball is cut into two hemispheres, the hemispheres cannot be leveled by bending alone. In theory, not even a very small piece of a sphere can be made to lie exactly flat by bending alone.

Summary

The experimental methods discussed here are all easy to introduce and teach. Having students use these methods on some simple surfaces is a good way to bring some lively activity into the classroom and to show the students that measurement in general is not as simple as one might at first think. More information on surfaces can be found in chapter 7 of *Mathematics: Its Content, Methods, and Meaning,* by A. D. Aleksandrov, A. N. Kolmogorov, and M. A. Lavrent'ev (Cambridge, Mass.: M.I.T. Press, 1963).

12

Measurement Activities for Exploring the Natural Environment

Marshall Gordon

The natural environment has been used and abused, on a large scale, since the Industrial Revolution. Significant implications may result from making the natural environment a part of the educational experience and helping students arrive at good approximations based on sound mathematics in their study of the environment. Students will have experienced mathematics by *doing* mathematics, and the facts, concepts, and values they explore and develop should lead both to a greater concern for the environment and to a greater appreciation for the role of mathematics in studying it. Therefore, some activities for exploring the natural environment will be examined: measuring the height of a forest canopy, the area of a field, and the area of a pond.

Measuring the Height of a Forest Canopy

Materials needed: a plastic straw, a 3 × 5 index card, a piece of string or fish line, a small weight, a trigonometry table (one-sheet version), some tape and rope, a small protractor, a meterstick

Figure 12.1 shows the angle clinometer, an instrument used to measure angles. Make certain that the straw is taped to the edge of the card, the string is centered on the straight edge of the protractor, and the protractor is parallel to the edge of the 3 × 5 card. If there are not enough protractors to go around, one can be made from cardboard. A trigonometry table can be attached to the back of the index card for accessibility.

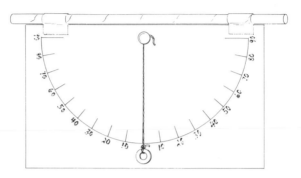

Fig. 12.1

When we look at the treetops through one end of the straw, the weight on the string naturally comes to rest perpendicular to the ground. The intersection of the string with a mark on the protractor indicates the measure of the angle that is made between the line of sight to the object and the line of sight directly in front of the viewer.

Sometimes when we try to measure the height of a pine forest's canopy, the foliage may be so dense that we "cannot see the trees for the forest." In this situation, we must find the heights of a few trees and take an average to determine the approximate height of the canopy.

Three lines of sight can be determined with the angle clinometer: to the base of the tree, straight ahead, and to the top of the tree (fig. 12.2). The

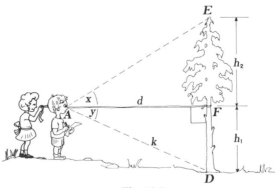

Fig. 12.2

length d could be determined by tying a rope around the tree and measuring the rope's length to the point where the viewer can see the entire tree. The angle measures x and y are found by looking through the straw to the top and to the base of the tree and reading the points of intersection of the string with the protractor. (It has been assumed that the maximum height of the tree is directly above the base.) The height h_2 can be determined from the relationship that $\tan x = h_2/d$, so that $h_2 = d \cdot \tan x$. The height h_1 can be determined using angle measure y or by measuring with a meterstick the height to where the rope is tied around the tree.

If h_1 is too high to permit a rope to be placed around the tree, then length k may be determined by tying the rope around the tree's base. Then $h_1 = k \cdot \sin y$, and $d = k \cdot \cos y$. Finally, $h_2 = d \cdot \tan x$. The height of the tree is then

$$h_1 + h_2 = k \cdot \sin y + d \cdot \tan x.$$

Note that the height of the observer is taken into account in this instance by the measures of x and y: the taller the person, the smaller the value of x and the larger the value of y.

Discussion

The reason the angle clinometer determines the angle measure can be seen by examining figure 12.3. Point O is where 0 will be on the protractor, and point B is where the weight attached to the string is pointed. If $m \angle BAO = 30°$, then since $m \angle OAE = 90°$, drawing $\overleftrightarrow{AF} \parallel \overleftrightarrow{BD}$ determines that $m \angle OAF = 60°$, and thus $m \angle FAE = 30°$.

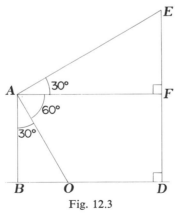

Fig. 12.3

Student projects

Determine the heights of various objects, such as the school, some trees, or houses. Note that the height of the viewer must be considered if the object and the viewer are on the same horizontal plane.

Devise a method to determine the average growth rate for different species of trees.

Measuring a Field

Materials needed: string (preferably marked off in meters), stakes, and a compass used in geometric constructions

Consider a field *ABCD* (fig. 12.4). Suppose by using the string and stakes we determined the dimensions to be $AB = 16$ m, $BC = 17.5$ m, $CD = 20$ m, $DA = 30$ m, and $BD = 23$ m. Then, with a scale ratio of 5 m to 1 cm, a scale drawing would show that $AB = 3.2$ cm, $BC = 3.5$ cm, $CD = 4$ cm, $DA = 6$ cm, and $BD = 4.6$ cm.

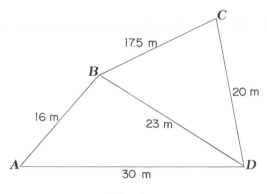

Fig. 12.4

In order to draw the scale diagram, a compass may be used to determine the points of intersection of the line segments. After $\overline{AD}$ is drawn to scale, locate the point of the compass at A and draw an arc with radius 3.2 cm. Then, with center D, draw an arc with radius 4.6 cm. Point B is located at the intersection of the two arcs: 3.2 cm from A and 4.6 cm from D. The same procedure can be used to locate point C.

To see if the scale drawing represents the original region proportionately, the method of trilateration can be used. In trilateration, additional segments are created on the scale diagram, and the scale lengths are determined. Then, these measures are compared to the corresponding real lengths in the field.

The method can be used as follows. Segments XY and WZ are drawn parallel to segments AD and BC, respectively (fig. 12.5). Point X is the midpoint of $\overline{AB}$, and Y is the midpoint of $\overline{BD}$. Point W is one-quarter the distance from B to D, and Z is one-quarter the distance from C to D. For example, XY should be one-half AD in the scale drawing and should

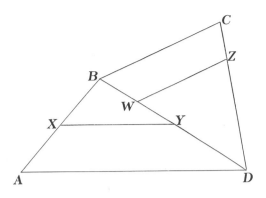

Fig. 12.5

measure 15 m in the field. If this is not found to be so, then the original measurements and the locations of the new points must be checked.

The area of region $ABCD$ can be found by determining the sum of the areas of the nonoverlapping triangular regions ABD and CBD. After the scale-drawing dimensions have been determined, they must be converted to the field measurements to determine the area of region $ABCD$. For instance, $BE = 2.4$ cm and $CF = 2.9$ cm, and so the area of region $ABD = \frac{1}{2}(2.4 \text{ cm})(6 \text{ cm}) = 7.2 \text{ cm}^2$, and the area of region $BCD = \frac{1}{2}(2.9 \text{ cm})(4.6 \text{ cm}) = 6.67 \text{ cm}^2$. Thus, the area of field region $ABCD = 7.2(25 \text{ m}^2) + 6.67(25 \text{ m}^2) = 346.75 \text{ m}^2$ (fig. 12.6).

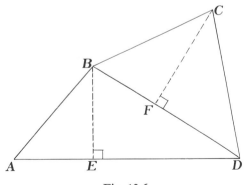

Fig. 12.6

Discussion

When a scale drawing is created so that m meters of a region are represented by c centimeters on a scale drawing, a convenient scale should be chosen. For example, if the maximum distance to be converted to scale is 30 m and this length is to be represented on an $8\frac{1}{2}$-by-11-inch piece of paper (approximately 21.6 cm × 28 cm), then a convenient scale would be 1.5 m to 1 cm. Although any scale can be used, care should be taken to devise one that gives a good visual representation of the region.

Student projects

Create scale drawings of regions encompassing a number of different trees. Locate the trees on the scale diagram. What hypotheses can be made about their growth patterns, root structure, and foliage?

Make a scale drawing of the classroom, including the students' and teacher's desks and an area for the doorway and the blackboard. Use the method of trilateration to check the scale drawing.

Measuring a Pond

Materials needed: stakes, string marked in meters

Suppose the dimensions of a pond require a rectangle 84 m by 48 m to encompass the total region of the pond. Then, choosing a scale ratio of 12 m to 1 cm would determine a scale drawing requiring a space of 7 cm by 4 cm. Figure 12.7 is a scale drawing showing intervals of 5-m distances.

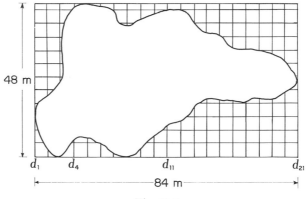

Fig. 12.7

At each interval measure, the distance from the rectangle to the pond can be determined. Then, with string and stakes a collection of k distances $(D_1, D_2, \ldots, D_k)$ from the rectangle to the pond can be found. When converted to scale, these measures $(d_1, d_2, d_3, \ldots, d_k)$ would represent the pond's curvature. For example, if the eleventh measurement D_{11} equals 10.8 m, then on the scale drawing, d_{11} would equal 0.9 cm. If a smaller interval width is chosen, a truer representation of the pond's curvature could be obtained.

In order to determine the area of the pond, two approximations can be used: either (1) average length times maximum width or (2) maximum length times average width. To find the average length of the pond, subtract d_i and d_j from the corresponding length of the rectangle, as shown in figure 12.8. This would be done for each interval both in the length and

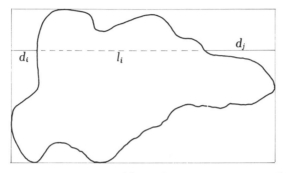

Fig. 12.8

in the width. It would provide a number of scale measures, as shown in table 12.1.

TABLE 12.1
LENGTH AND WIDTH APPROXIMATIONS OF THE POND

Length (cm)	Width (cm)
1.50	0.65
2.80	1.15
5.45	1.30
5.65	1.25
5.80	2.10
5.80	1.80
3.65	2.25
3.70	2.40
3.50	2.70
3.20	2.70
2.00	3.00
43.05	3.05
	3.00
	3.20
	3.45
	2.30
	1.20
	37.50

The seventeen width measurements have an average width of 2.21 cm, and the eleven length measures yield an average length of 3.91 cm. To determine the maximum width and length of the pond, a rectangle is circumscribed around the pond. Figure 12.9 shows that the maximum length is 7 cm and the maximum width is 4 cm. There are two approximations of the pond's area:

1. Maximum length times average width $= (7 \times 12)\ (2.21 \times 12) =$ 2227.68 m²
2. Maximum width times average length $= (4 \times 12)\ (3.91 \times 12) =$ 2252.16 m²

Since the average width was determined from seventeen measurements and the average length from eleven, the first approximation should be closer to the true area. Generally, in approximating the area of a closed curve, the greater the number of intervals, the closer the approximation to the area.

A grid approximation can also be used to approximate the area. Figure 12.9 shows the 448-squares area obtained by placing the scale drawing of the 28-cm² area on a grid 0.25 cm by 0.25 cm. Counting shows approximately 202 squares outside the pond region, so that the scale drawing of the pond contains approximately 246 squares. In figure 12.9, each scale-

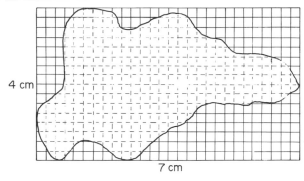

4 cm

7 cm

Fig. 12.9

drawing square represents an area of 3 m by 3 m, or 9 m². Hence, when this grid approximation is used, the area of the irregularly shaped region is 246 by 9 m², or 2214 m². The determination of the number of squares within the closed region depends on the curvature of the shape, since the portion of the square regions on the inner perimeter varies with the degree of curvature. For example, if a scale ratio of 4 m to 1 cm had been used instead of 12 m to 1 cm, the grid method would have yielded a better approximation, since it would have represented the curvature more accurately. However, the measurements produced by these two methods differ by less than 2 percent.

Discussion

Although the "averaging rectangle" approximation is derived from calculus, students can obtain an intuitive understanding of the process by examining the area of a simple region (fig. 12.10).

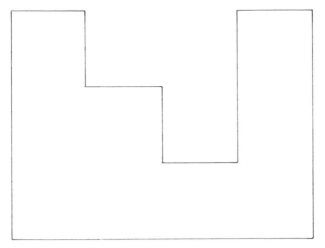

Fig. 12.10

Partition this region into four rectangles so that the area is $b_1h_1 + b_2h_2 + b_3h_3 + b_4h_4$ (fig. 12.11). Each of the bases has equal length, b

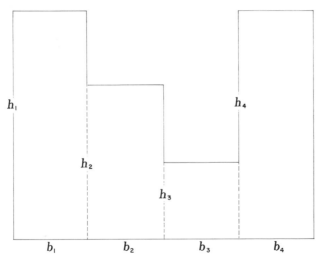

Fig. 12.11

units; therefore, the area is $b(h_1 + h_2 + h_3 + h_4)$. The sum of the heights is $4\bar{h}$, where $\bar{h}$ is the arithmetic mean of the heights; thus, the area is $b(4\bar{h})$. Notice that "$4b$" is the total length of the base, that is, it is the maximum base. Thus, this area can be written as

$$\text{maximum base } (4b) \times \text{average height } (\bar{h}).$$

The total area is then equal to the area of a rectangle whose base is the maximum base and whose height is the average of the four rectangular heights. For example, if each rectangle had a base of length 2 cm and heights of 6, 4, 2, and 6 cm, then the total area would be the sum of the areas of the rectangles, 36 cm². The "averaging rectangle" would have a base of 8 cm and an average height of 4.5 cm, and thus an area of 36 cm².

Two familiar geometric shapes, the trapezoid and the triangle, will now be considered, and their areas in terms of an "averaging rectangle" will be interpreted. Recall that the area of a trapezoid is $\frac{1}{2}h(b_1 + b_2)$. The height, h, is the maximum height, and $\frac{1}{2}(b_1 + b_2)$ is the average of the two bases. This seems to agree with the principle expressed so far, except that "$\frac{1}{2}(b_1 + b_2)$" represents the average of only two horizontal segments. Is the average of the lengths of the first (top) and the last (bottom) line segments equal to the average of the lengths of a collection of horizontal line segments drawn at equal intervals within the trapezoid? The answer is yes! Therefore, taking the average of the lengths of the first and the last segments will yield the same value as taking the average of any number of horizontal segments (partition segments) drawn at equal intervals to the bases.

The proof that the averages are the same lies in the fact that the sides of the trapezoid are line segments. When segments of equal length are cut off, similar triangles are formed, and thus the lengths of the segments, k lengths, form an arithmetic sequence: $n, n + (a + b), n + 2(a + b), \ldots,$ $n + (k - 1)(a + b)$, where the common difference is $(a + b)$. Figure 12.12 demonstrates the relationship. The perpendiculars dropped to the

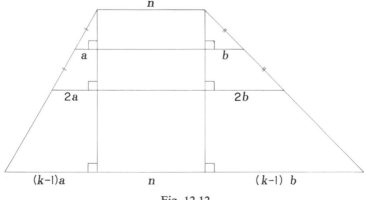

Fig. 12.12

larger base create similar triangles in the ratio of $a{:}2a{:}3a{:} \ldots {:}(k - 1)a$ on the left side and $b{:}2b{:}3b{:} \ldots {:}(k - 1)b$ on the right side. Since this is an arithmetic sequence, the sum of k terms is

$$S_k = \frac{k(a_1 + a_k)}{2}$$

Hence, the average of the k terms is found by dividing the sum by the number of terms, k; that is,

$$\frac{S_k}{k} = \frac{(a_1 + a_k)}{2}$$

the average of the first and last terms. Therefore, an "averaging rectangle" can be constructed with height h and the average of the lengths of the bases of the trapezoid as the base, and the areas would be the same.

For the triangle, the height of the triangle is the maximum height. Consider "$\frac{1}{2}b$" as representing the average of the sum of the top base (vertex with length 0) and the bottom base, b. The average is $\frac{1}{2}(b + 0) = \frac{1}{2}b$. Hence an "averaging rectangle" would be constructed with its height the maximum height of the triangle, its base one-half the base of the triangle, and the areas would be the same.

When the area of a closed curve such as a pond is approximated by determining an "averaging rectangle," an approximation rather than the exact area will always result. The approximations will be close to the real area as long as the number of equal intervals created by the partitioning is not too small. In calculus, the average value of a function, average height, is determined by the formula

$$\bar{y} = \frac{1}{b - a} \times \text{area (under the curve)}.$$

In this situation, the area is determined by the product of the maximum base and the average height (fig. 12.13):

$$\text{area} = \bar{y}(b - a)$$

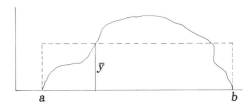

Fig. 12.13

Student projects

In the classroom, take a rope about five meters long and create an irregular closed shape on the floor, like a pond. Determine a scale ratio and make a scale drawing of this shape. Approximate the area of this region.

Make a scale drawing of an irregular shape on tagboard or heavy construction paper. Is there a different way to approximate the area of a region enclosed by an irregular shape, perhaps by using weighing scales and scissors?

What argument would explain that the formula "average width times average length" would not yield a good approximation of the area of a closed curve?

13

Mathematical Analyses of Physical Measurements: Measuring the Velocity of a Projectile

Guy Schupp

This essay describes a simple experiment involving two of the laws most cherished by physicists: the conservation of energy and the conservation of momentum. Before the experiment and its mathematical analysis are discussed, however, the overall problem of physically measuring some quantity like a length or an area and the inherent uncertainties associated with the particular measurements should be considered.

As "searchers for truth," teachers may want to obtain the "true" length of a table top or desk. If students were asked to measure the length of a table two meters long (2 m), what results would be expected, and how would they be related to the true length? If the students were asked simply to measure the length to the nearest centimeter (cm), everyone might agree that it measured 200 cm. If the students were asked to measure the length to the nearest 0.5 millimeter (mm), however, most of us would be surprised if everyone got the same answer, even though the same table was being measured with the same meterstick! Why? Simply because it is difficult to determine the ends of the table exactly; because the ends may not be parallel; because it is difficult to move the meterstick and still be exact to less than 0.5 mm; because. . . . Uncertainties arise when any physical measurement is pushed far enough, even when the greatest care is taken.

These inherent uncertainties are often called errors and may be classified as illegitimate, systematic, and experimental. Illegitimate errors may be the result of "goofs" in reading the meterstick or computational mistakes in more complicated experiments. Systematic errors may arise either from using an old meterstick that has shrunk with age or from consistently tilting one's head when making a reading and thereby introducing a parallax. Experimental errors involve some of the problems mentioned earlier concerning the ends of the table as well as any psychological tendencies for estimating that smallest fraction of a millimeter differently on subsequent remeasurements.

Calculation of Experimental Errors

Standard deviation

Consider the results that could be expected from n independent measurements of the table's length, x. With equal care taken in each of the measurements, a simple average of the length x, defined by $x = (x_1 + x_2 + x_3 + \ldots + x_n)/n$, would give a "best" value for the length. Eliminating illegitimate and systematic errors means that the variation in the individual measurements will be related only to experimental error. A common measure of experimental error is the standard deviation, which can be calculated in a straightforward manner. For the case of n independent measurements of x, the standard deviations σ_x (of an individual measurement of x) or $\sigma_{\bar{x}}$ (of the average value $\bar{x}$) are given by

$$\sigma_x = \sqrt{\frac{d_1^2 + d_2^2 + \ldots + d_n^2}{n-1}} \text{ and } \sigma_{\bar{x}} = \sqrt{\frac{\sigma_x^2}{n}}, \qquad (1)$$

where $d_1 = x_1 - \bar{x}$, $d_2 = x_2 - \bar{x}$, and so on. The quantity $n - 1$ occurs in the expression for σ_x because it represents the number of independent measurements remaining, since $\bar{x}$ requires that $d_1 + d_2 + \ldots + d_n = 0$. The value for $\bar{x}$ given above is called *best* because it minimizes the value of $\sigma_{\bar{x}}$. It is customary to give an experimental result as the average plus or minus the standard deviation on the average, that is, $\bar{x} \pm \sigma_{\bar{x}}$.

If the length of the table were measured by 100 different people, the number of measurements within 0.5-mm intervals might look like the histogram with dashed lines shown in figure 13.1. The average value in

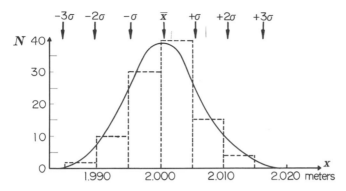

Fig. 13.1. A histogram (dashed lines) of the data and the error, or Gaussian, curves (solid line) for a normal distribution that has been normalized to represent 100 measurements. N is the number of measurements per 0.5 mm, and x is the measured length of the table.

this example is given by $[2(1.987\ 5) + 10(1.992\ 5) + 30(1.997\ 5) + 40(2.002\ 5) + 15(2.007\ 5) + 3(2.012\ 5)]/100 = 2.000\ 75$. In the limiting case of a very large number of measurements and very small intervals, this histogram approaches what is called a "normal" distribution (also called an error, or Gaussian, curve), given by $\exp[-(x - \bar{x})^2/2\sigma_x]$. A plot of this curve is shown by the solid curve in figure 13.1, with $\bar{x} = 2.000\ 75$, $\sigma_x = 0.005\ 09$, and the area normalized to represent 100 measurements. The fraction of the area of the normal distribution between $-\sigma$ and $+\sigma$ is 68.3 percent—meaning that in our example sixty-eight measurements would statistically have fallen in the interval from 1.9957 to 2.0058. For completeness, the fractional area between -2σ and $+2\sigma$ is 95.6 percent and that between -3σ and $+3\sigma$ is 99.7 percent.

Propagation of errors

Although the previous example illustrates how a standard deviation can be calculated from several measurements of the same quantity, more mathematically and physically interesting situations arise when many different experimentally measured quantities need to be combined to give a final result. A simple example would be to determine the area of the table. If the best values for the length and width are 200.0 ± 0.5 cm and 100.0 ± 0.4 cm, respectively, what will the uncertainty in the area be? Expanding $(200.0 \pm 0.5) \times (100.0 \pm 0.4)$ cm^2 gives $(20\ 000 \pm 80 \pm 50 +$ higher-order terms) cm^2, but it does not indicate what to do with the $\pm$'s. If the $+$'s for the length and the width go together, the errors are said to be "dependent," and the uncertainty in this event would be ± 130 cm^2. In measurements of this type, however, a $+$ in the length measurement is usually associated with a $+$ or a $-$ in the width measurement with equal probability. In this event, the errors are said to be "independent." For these situations, the theory of error gives a prescription for calculating the standard deviation, σ_V, for a quantity V that may depend on any number of independent variables (see Beers [1953]).

Typically, in many experiments, several quantities x, y, z, . . . will be measured independently, which may have experimental uncertainties σ_x, σ_y, σ_z, . . . , and must be combined in some specific way to give a particular quantity, say, V. Usually these σ's will have to be estimated according to how closely one can read a scale, a meter, some coordinates, and so on, and will not always be determined from deviations about an average. It could then be said that V depends on x, y, z, . . . , or $V = V(x, y, z, \ldots)$. Theory then gives the following, rather complicated expression for σ_V in terms of the σ_x, σ_y, and so on:

$$\sigma_V = \sqrt{\left(\frac{\partial V}{\partial x}\,\sigma_x\right)^2 + \left(\frac{\partial V}{\partial y}\,\sigma_y\right)^2 + \left(\frac{\partial V}{\partial z}\,\sigma_z\right)^2 + \cdots} \quad (2)$$

The expression $\partial V/\partial x$ is a partial derivative and signifies taking the derivative of V with respect to x while holding all the other variables $(y, z, \ldots)$ constant. Hence, it is the rate of change of V with respect to x.

Several important and interesting examples pertaining to σ_V follow:

Example 1. Let $V(x, y, z) = x + y - z$; then

$$\sigma_V = \sqrt{(1\sigma_x)^2 + (1\sigma_y)^2 + (-1\sigma_z)^2} = \sqrt{\sigma_x^2 + \sigma_y^2 + \sigma_z^2}.$$

This simple result indicates that in addition or subtraction the independent errors combine as the square root of the sum of the squares.

Example 2. Let $V(x, y, z) = xy/z$; then the expression for σ_V may be simplified by dividing by V to give

$$\frac{\sigma_V}{V} = \sqrt{\left(\frac{\sigma_x}{x}\right)^2 + \left(\frac{\sigma_y}{y}\right)^2 + \left(\frac{\sigma_z}{z}\right)^2},$$

which indicates that in multiplication or division the fractional errors σ_x/x, and so on, combine as the root of the sum of squares to give the fractional error σ_V/V. Since most laboratory uncertainties are of a small percentage, we could multiply all the errors by 100 and then deal with percentage errors. For example, if $\sigma_x/x = 2\%$, $\sigma_y/y = 2\%$, $\sigma_z/z = 3\%$, then

$$\frac{\sigma_V}{V} = \sqrt{2^2 + 2^2 + 3^2} = \sqrt{17} \cong 4\%.$$

Example 3. Let $V(x) = x^n$; then $\sigma_V = \sqrt{(nx^{n-1}\sigma_x)^2} = nx^{n-1}\sigma_x$,

or $\sigma_V/V = n(\sigma_x/x)$, and we see that the fractional or percentage error gets multiplied by the exponent and is doubled in squaring or halved in taking the square root. Notice that if $V(x)$ were considered to be $x \cdot x$, the result of example 2 would lead to $\sigma_V/V = \sqrt{2}(\sigma_x/x)$ rather than to the correct result, $2(\sigma_x/x)$. The reason for this difference is that in applying the incorrect result, we would be assuming that the σ_x's on the x's in $x \cdot x$ are independent, when in fact they are not.

Example 4. A final interesting case shows that the error on V may depend strongly on its value. Let $V(x) = \cos x$; then

$$\sigma_V = \sqrt{(-(\sin x)\, \sigma_x)^2} = (\sin x)\, \sigma_x.$$

Consider the following values for $x \pm \sigma_x$:

a) $\pi/2 \pm 0.1$ rad; then $V \pm \sigma_V = 0.0 \pm 0.1$

b) 0 ± 0.1 rad; then $V \pm \sigma_V = 1.0 \pm 0.0$

These results can be easily understood graphically, as shown in figure 13.2.

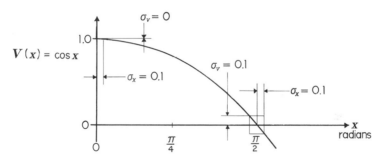

Fig. 13.2. A plot of cos x versus x showing how the uncertainty in cos x, arising from an uncertainty in x, is dependent on the x value.

Returning to the determination of the area of the table, we can directly apply example 2 to give

$$\frac{\sigma_{\text{area}}}{\text{area}} = \sqrt{\left(\frac{0.5}{200}\right)^2 + \left(\frac{0.4}{100}\right)^2} = 4.7 \times 10^{-3},$$

or

$$\sigma_{\text{area}} = (2.0 \times 10^4)(4.7 \times 10^{-3})\ \text{cm}^2 = 94\ \text{cm}^2.$$

Another way to consider this result in terms of the ± 80 and ± 50 discussed earlier is to understand that $94 = \sqrt{80^2 + 50^2}$, which means that the independent errors are added quadratically.

The Experiment

This prologue has raised questions that are embedded in every mathematics problem involving measurement. Such issues must be recognized if the mathematical analysis of the problem is to be interpreted properly. Let us now consider an experiment that uses concepts from both mathematics and physics—an experiment that is clearly a measurement problem, the solution of which will allow us to determine the velocity of a projectile.

Equipment and procedures

The velocity of a common suction-cup dart as it emerges from a toy spring-dart gun can be measured by using two of the most important laws of physics: the conservation of energy and the conservation of momentum. The dart is fired so that it sticks to a massive block suspended from some fixed object by strings like a long pendulum. Following the impact, the combined block and dart will swing until it rises to a maximum height that can be related to the velocity of the dart. Operationally, the rise in height of the pendulum can best be measured by the horizontal distance that the pendulum swings. This problem is commonly referred to in physics textbooks as the ballistic pendulum (see Halliday and Resnick [1970]).

The following materials are needed: toy spring-dart gun, dart with rubber suction cup, target block with smooth end, meterstick, balance, string, and weights. The target block can easily be made from a block of wood approximately 6 cm by 6 cm by 12 cm, with a piece of glass or Formica glued to one end so that the dart will stick to it. The block is suspended about two meters from the ceiling (or from some appropriate support) by strings, which are looped under screws on the side of the block near the ends, as shown in figure 13.3. To keep the block and dart horizontal, the strings must be arranged so that they remain parallel when the pendulum swings. A meterstick is clamped in a horizontal position immediately below the target block, as seen in figure 13.3. A light-weight metal slider with an upward-projecting wing about 1 cm high is set on the meterstick so that the rear end of the target block just touches the wing when the pendulum is at rest. Record the position of the slider. When ready, hold the loaded gun in a horizontal position a few centimeters from the smooth front end of the block and fire! The gun should be held close enough to insure a solid hit, but far enough away so that the dart leaves the muzzle before striking the block. The block and dart will swing like a pendulum, pushing the slider along the meterstick. Without resetting the slider, refire the dart until the slider no longer moves. This method for measuring the horizontal swing of the pendulum will minimize frictional effects on the slider and give a value for the maximum velocity of the dart. Record the final position of the slider and recheck its original position.

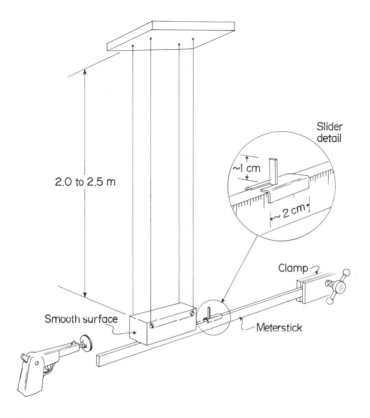

Fig. 13.3. Experiment for measuring the velocity of a rubber-tipped dart. The length of the strings must be adjusted so that the block swings horizontally. The meterstick should be clamped in position about 2 mm below the target block. The slider must move freely on the meterstick.

The difference between these positions will be the distance d referred to later in equation (5). Use the balance to determine the masses of the dart and block; carefully measure the length of the supporting strings.

Mathematical analyses

The velocity of the dart can then be computed as follows: Let

m = mass of the dart with suction cup
M = mass of the target block (including all screws and one-half the mass of the strings)
v = velocity of the dart before impact (this is what is sought)
V = velocity of block and dart immediately after impact

Photographs of the experimental arrangement depicted in figure 13.3. The target block was actually constructed with aluminum sides and with the smooth surface midway between the supporting strings for slightly better stability.

Conservation of momentum

The momentum of an object is simply the product of its mass and velocity. In any collision where no net external forces act, the momentum of the system is conserved (does not change). Here the system refers to the dart and the target pendulum. Therefore, we can equate the momentum of the system immediately before and after the collision by the equation

$$mv = (m + M)V. \tag{3}$$

Since m and M have been measured, V must be determined to calculate v.

Conservation of energy

Conservation of energy, as it applies to the target pendulum and the dart after the collision, is used in determining V. Although a lot of mechanical energy is converted to heat and sound energy during the impact, the mechanical energy of the pendulum is conserved after the collision. The mechanical energy of an object is the sum of its kinetic, or motional, energy and its potential, or positional, energy. The mechanical energy of the pendulum immediately after impact is completely kinetic, and at the top of its swing it is completely potential. The kinetic energy of an object is given by the product of one-half its mass and the square of its velocity. Near the surface of the earth the potential energy of a body is given by the product of its mass, height, and acceleration of gravity. The symbol g is used to represent the gravitational acceleration, which has a very nearly constant value of 980 cm/s² in the United States. (An empirical value for g at any place on the surface of the earth is given by $g = (980.616 - 2.5928 \cos 2\phi + 0.0069 \cos^2 2\phi - 3.086 \times 10^{-6}H)$ cm/s², where ϕ is the latitude and H is the elevation above sea level in centimeters.) Equating the kinetic energy of the pendulum immediately after the collision to the increase in potential energy at the top of the swing gives

$$\tfrac{1}{2}(m + M)V^2 = (m + M)gh, \tag{4}$$

where h is the increase in the vertical height of the pendulum at the top of its swing. It should be noted here that if the block and dart did not remain horizontal during the swing, the value of h would not be clearly defined. Equation (4) can be solved for $V = \sqrt{2gh}$, but as mentioned earlier, it is better to obtain h from the horizontal swing, since the vertical rise is less than a centimeter and would be very difficult to measure directly. Figure 13.4 shows the relationship between the length of the pendulum strings l, the height that the block and dart are raised h, and the horizontal distance through which the slider on the meterstick is pushed d. The Pythagorean theorem gives

$$l^2 = (l - h)^2 + d^2$$

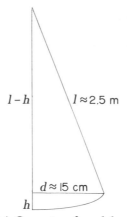

Fig. 13.4. Geometry of pendulum swing

or

$$2lh(1 - h/2l) = d^2.$$

Typically, $h/2l$ is 0.5 cm/500 cm, or 0.001, and can be neglected with respect to other experimental uncertainties to give

$$h = \frac{d^2}{2l}.$$

Substitution for h into equation (4) and thence for V into equation (3) gives the desired expression for the velocity of the dart:

$$v = (1 + M/m) \, d \sqrt{\frac{g}{l}} \tag{5}$$

Equation (5) shows the explicit dependence of the velocity of the dart on the measured quantities m, M, d, and l. Since the M/m ratio is dimensionless, the dimensions for v will be in cm/s if d is measured in cm, g in cm/s^2, and l in cm. Typical values for the experimental quantities are $m = 5.0 \pm 0.1$ gm, $M = 200.0 \pm 0.1$ gm, $d = 14.0 \pm 0.1$ cm, $l = 250.0 \pm 0.2$ cm, and $g = 980 \pm 1$ cm/s^2, which give a velocity of 1137 cm/s (or 25.4 miles per hour).

Experimental uncertainties

An uncertainty assignment for the value of v can be determined in either of the two ways discussed earlier. If many different determinations have been made, σ_v can be determined from deviations about the average by using equation (1). From a more fundamental point of view, however, the uncertainty for v can be calculated by the method discussed in conjunction with finding the area of the table. In this situation, by using equation (2), we can show σ_v (by taking the prescribed partial derivatives) to be

$$\sigma_v = d \left\{ \frac{g}{l} \right\}^{1/2} \left\{ \left(\frac{M}{m} \frac{\sigma_M}{M} \right)^2 + \left(\frac{M}{m} \frac{\sigma_m}{m} \right)^2 + \left[\left(1 + \frac{M}{m} \right) \frac{\sigma_d}{d} \right]^2 \right.$$

$$\left. + \left[\left(1 + \frac{M}{m} \right) \frac{\sigma_g}{2g} \right]^2 + \left[\left(1 + \frac{M}{m} \right) \frac{\sigma_l}{2l} \right]^2 \right\}^{1/2} .$$

Since $\dfrac{M}{m} \cong 40$, replacing $\left(1 + \dfrac{M}{m} \right)$ by M/m gives the very good approximation

$$\frac{\sigma_v}{v} = \sqrt{ \left(\frac{\sigma_M}{M} \right)^2 + \left(\frac{\sigma_m}{m} \right)^2 + \left(\frac{\sigma_d}{d} \right)^2 + \left(\frac{\sigma_g}{2g} \right)^2 + \left(\frac{\sigma_l}{2l} \right)^2 } ,$$

where the factors of 2 in the denominators of the last two terms come from the square-root dependence in equation (5). Within this approximation it is easy to see how the fractional errors of the various quantities contribute to σ_v/v. Fractional errors calculated from the typical values are $\sigma_M/M = 0.0005$, $\sigma_m/m = 0.020$, $\sigma_d/d = 0.007$, $\sigma_g/g = 0.001$, and $\sigma_l/l = 0.0004$. When combined quadratically, only the uncertainties in m and d contribute substantially to the overall error assignment to give

$$\frac{\sigma_v}{v} \cong \sqrt{ (0.02)^2 + (0.007)^2 } = 0.021 = 2.1\% .$$

Although the value of v will depend on the particular gun used, an experimental uncertainty for v of 2 or 3 percent should be expected. The final result for $v \pm \sigma_v$ from the typical values would be 1137 ± 24 cm/s.

Experimental checks

Once the experiment and its error analysis have been completed, the question of how the results might be checked could be raised. This question leads naturally to a deeper appreciation of the ideas of energy and momentum conservation and possibly to more confidence in the results themselves. One method for checking the muzzle velocity of the dart is to fire it vertically and measure its maximum rise in height. In this experiment, the kinetic energy of the dart as it leaves the gun can be equated to its increase in potential energy at the top of its flight to give

$$\tfrac{1}{2}mv^2 = mgy, \text{ or } v = \sqrt{2gy},$$

where y is the maximum rise in height of the dart. Values of v measured in this manner are usually 3 to 5 percent smaller than the value calculated from equation (5) because of the air resistance to the dart. Alternatively, if the dart is fired horizontally at some height z above the floor, the time it takes for the dart to hit the floor is the same as if it had been simply dropped instead of fired. The velocity of the dart can then be related to the horizontal range, R_H, by the equation

$$v = R_H \sqrt{g/2z}.$$

A main difficulty in checking v in this manner is that the dart must be fired truly horizontally. The sights on some toy guns make this quite easy. Simply invert the gun and allow it to rest on its barrel sights near the edge of a horizontal surface before firing. The value for R_H should be the distance between the place where the suction cup strikes the floor and that point on the floor directly below the suction cup as it leaves the muzzle. The height z is measured from the lower edge of the suction cup to the floor when the gun is in the firing position. To be consistent with determining the maximum value for v, several horizontal firings should be made and the longest, R_H, used. Again, air resistance decreases the value of v measured in this manner by a small percentage.

Summary

In summary, the measurement of the velocity of a projectile as described here affords the opportunity to investigate techniques of measurement as well as different mathematical analyses. Comparisons between the different measurements lead students naturally to the considerations of experimental uncertainties and to the validity and application of physical laws.

REFERENCES

Beers, Yardley. *Introduction to the Theory of Error*. Reading, Mass.: Addison-Wesley Publishing Co., 1953.
Halliday, David, and Robert Resnick. *Fundamentals of Physics*. New York: John Wiley & Sons, 1970.

14

Manipulative Devices for Elementary School Measurement Activities

Robert L. Jackson
Glenn R. Prigge

This essay provides descriptions and illustrations of a variety of manipulative aids that could be used in teaching measurement. The few illustrative activities outlined for each device should, of course, suggest other possible activities. The devices are grouped under the following six headings: length, temperature, area, volume, mass and weight, and metric place-value aids.

At the end of the essay is a numbered list of companies that supply various devices designed expressly for elementary measuring activities. The numerals in parentheses at the end of the discussion for each device refer to those companies that can provide such a device, along with an approximate price. Thus (1, 14, 15 ≈ $4.50) shows that the companies numbered 1, 14, and 15 could supply the device for about $4.50.

Length

Centimeter or inch bars

These versatile bars or rods, available in both scored and nonscored form, are generally color coded for easy identification. Normally, a teacher uses unscored bars to develop number concepts by having children manipulate the bars through specially designed discovery or guided-discovery lessons that introduce children to the idea of a standard unit of measure. The unscored bars are particularly valuable as units for measuring the length of relatively small classroom objects, such as books, desk tops, or crayon boxes. Using different unscored bars as units demonstrates to children that a single object can have several measures and helps them understand the need for a standard unit like a centimeter or an inch.

The scored bars (fig. 14.1) are effective in working with area and volume problems. Children might be asked to cover an area with bars and

Fig. 14.1

count the number of "units" needed. If the children measure the length and width of the given square or rectangle with the chosen bar (fig. 14.2) and compare those measures to the number of units of area, the formula for determining area can be discovered. Similarly, children can pack specially designed boxes with scored bars and develop the notion of volume

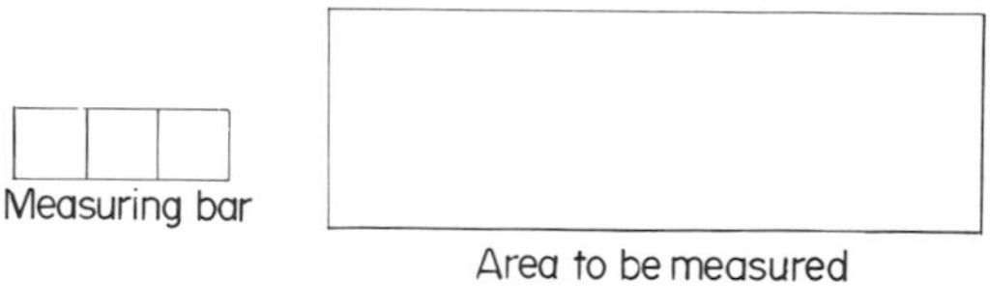

Measuring bar

Area to be measured

Fig. 14.2

or the unit cube. Bars can be used to develop the concept of seriation (fig. 14.1), thus preparing the child for "greater than" or "less than" relationships. This manipulative aid is a concrete way for the unit—the inch or the centimeter—to be introduced.

Bars usually come in sets, vary in length from one unit to ten units, and differ in color according to the distributor. Some of the bars are flat with no scorings, some are scored only on one side, and others have the scoring on four sides. (2, 3, 4, 6, 8 ≈ $3.00)

Rulers

After children have had experiences measuring and comparing the lengths of objects with nonstandard measuring devices, they can be introduced to the unitized ruler and can use it effectively in measurement activities.

Rulers are available in both inches and centimeters (fig. 14.3). Although no standard-length ruler has been determined best for elementary school children, many educators advocate a decimeter (10 cm) ruler for early experiences. Metric rulers are available in varying lengths, from ten centimeters to fifty centimeters, and can be obtained with millimeter, centimeter, and decimeter gradations. (1, 2, 4, 6, 7, 8, 9, 12, 13, 14, 15 ≈ $0.50)

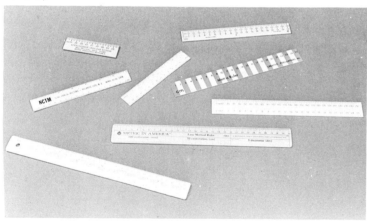

Fig. 14.3

Cloth tapes

Measuring tapes, which generally come in lengths of 150 centimeters, are especially useful for measuring relatively great lengths, such as the length and width of a classroom, the length of a corridor, or the length and width of a closet. Since the measurement of these distances may not be inherently interesting to children, such activities should have some specific purpose that would make them attractive and worthwhile. The longer tapes, since they do not require skill in the addition of several measures, are particularly valuable for young children. Some tapes are divided into decimeter sections of different colors, with a single color representing one decimeter (fig. 14.4). Such tapes help children sense the relationship between the decimeter and the centimeter. Similar tapes, marked in inches and feet, are also available. (1, 2, 4, 6, 8, 9, 12, 14 ≈ $0.50)

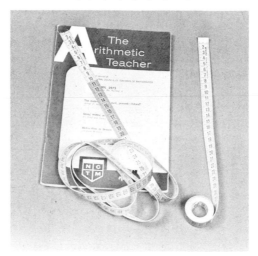

Fig. 14.4

Meterstick or yardstick

The meterstick or yardstick (fig. 14.5) is a rigid measuring instrument that is important for every child's learning experiences, since it helps develop the concept of a basic unit of length, the standard meter or yard. Laboratory lessons can be developed in which each child completes activities that use the meter or yard as the standard unit.

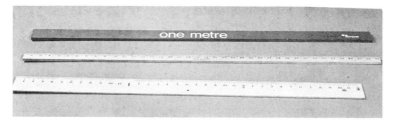

Fig. 14.5

The meterstick will help the child see that 100 centimeters = 10 decimeters = 1 meter. The yardstick enables the child to see similar relationships among inches, feet, and yards; the fact that these units of measure cannot be related to each other in a simple decimal way establishes one major advantage of the metric system. (1, 2, 4, 6, 7, 8, 9, 12, 13, 14 ≈ $3.00)

Metric number line

One way to bring the metric system into the classroom is to use number lines calibrated in metric units (fig. 14.6). The number line could be used on each child's desk and calibrated in centimeters, with the demonstration line in decimeters. Using these lines for arithmetic operations will provide each child with metric-unit experiences. (16 ≈ $2.00)

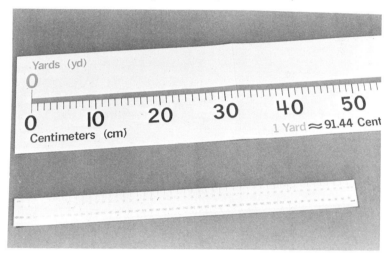

Fig. 14.6

Metric trundle wheel

The trundle wheel is an interesting device for children to use to determine the perimeter of irregularly shaped regions (fig. 14.7). Metric trundle

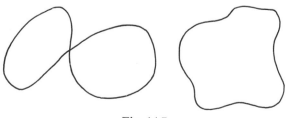

Fig. 14.7

wheels (fig. 14.8) usually have gradations in centimeters and decimeters and are so constructed that the circumference of the wheel is precisely one meter. A hard plastic handle is attached to the wheel for maneuverability. The child places the zero of the scale on the starting point and then rolls the wheel along the curve. A clicker indicates when a complete revolution, one meter, has been traveled. As the wheel is rolled to the end of the curve, the measurer counts the clicks, takes the reading from the trundle wheel at the end of the path, and adds to it the number of clicks (meters) counted.

Fig. 14.8

Nonmetric trundle wheels have gradations in inches and feet and a circumference of one yard.

Small trundle wheels, for use at a child's desk, can be made from posterboard circles having a circumference of one decimeter. The upper grades can use the trundle wheel to determine the value of π: circumferences can be measured with the wheel and the numbers obtained compared with diameters of circles. (1, 2, 4, 5, 6, 8, 9, 11, 12, 13, 14 ≈ $7.00)

Metric calipers

Calipers are constructed primarily to determine inside and outside diameters of cylinders. The caliper shown in figure 14.9 consists of two sets

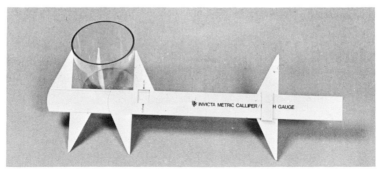

Fig. 14.9

of adjustable arms joined to a plastic frame. To find the outside diameter of a cylinder, the caliper is set as shown and the measure compared to a ruler. Some calipers have built-in rulers so that the measure can be read directly from the caliper. These devices can be used with the trundle wheel to establish the value of π.

Children could also measure the distance between two parallel lines and discover the theorem that says the distance between two parallel lines is the perpendicular distance between these two lines. (1, 2, 4, 5, 6, 8, 9, 11, 13, 14 ≈ $5.00)

Bow calipers

The bow caliper is a versatile instrument that measures in either centimeters or angular degrees. The device shown in figure 14.10 will measure up to seventy centimeters. Some bow calipers are designed to be used as a compass for geometric constructions.

Bow calipers have two curved arms that wrap around the object being measured; the reading is taken from a slide located on the caliper. Measuring circular and spherical objects is an excellent classroom activity that helps children learn how to measure diameters of cylinders and spheres. (1, 2, 4, 5, 8, 9, 11, 13, 14 ≈ $5.00)

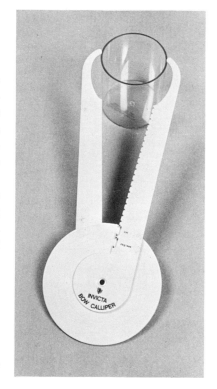

Fig. 14.10

Height measure

The height-measuring instrument enables children to measure their heights at established intervals throughout the school year. It also can be the start of many measurement activities that require recording and record keeping, for example, the growth of a bean or corn plant. Children can keep weekly or monthly growth records and construct statistical summaries and graphs.

Some height-measuring instruments are mounted on the wall; others are freestanding (fig. 14.11). (1, 4, 8 ≈ $14.00)

Fig. 14.11

Foot measure

This device can be used to determine foot sizes in either the Imperial or the metric system. The instrument shown in figure 14.12 has both calibrations.

Exercises in recording data can help children recognize relationships between the two systems of measure. Such activities avoid the rather complex process of determining equivalences through arithmetical operations. (4, 12 ≈ $9.00)

Fig. 14.12

Metal micrometer

The micrometer is used to measure thicknesses of materials with great precision. The maximum thickness that can be measured with the micrometer shown in figure 14.13 is one inch. Learning how to use the scale on a micrometer is an excellent exercise for older elementary school children. The scales provide opportunities for children to read gradations on the

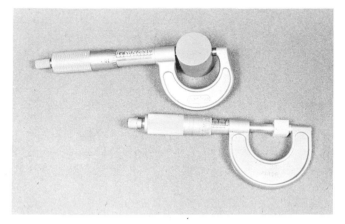

Fig. 14.13

micrometer's barrel to the nearest 0.025 inch and on the thimble to the nearest 0.001 inch. If a metric micrometer is used, the barrel gradations can be read to the nearest 0.01 millimeter. (1, 2, 6, 8, 11, 12, 13, 14 ≈ $6.00)

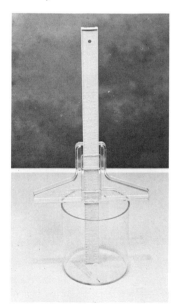

Depth gauge

The depth gauge (fig. 14.14) is used to measure the internal depth of a container and can be purchased with either the metric or the customary scale. Children can compare the depths of several different containers. (2, 6, 8, 9, 11, 13, 14 ≈ $8.00)

Fig. 14.14

Clinometer

A clinometer can be used to measure the angle of inclination between the ground and an object above ground level, thus enabling children to find the height of a tree, mountain, or building.

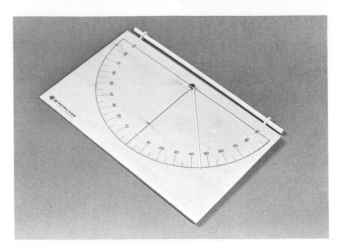

Fig. 14.15

This instrument is basically a 180° protractor (with one-degree gradations from 90° to 0° to 90°) with a sighting tube and a pendulum to
indicate the measure of the angle of the object sighted in relation to the
horizontal (fig. 14.15). With this angle and the measure from the viewer
to the base of an object, the child can construct a scale triangle to determine the height of the object (fig. 14.16).

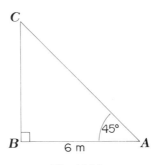

Fig. 14.16

In triangle *ABC, BC* is the height of a tree to be determined. The child
measures side *AB* and finds it to be 6 m. The angle of inclination read from
the clinometer is 45°. A simple scale drawing would enable the child to
determine *BC* (6 m). A table of trigonometric ratios might also be used.
(2, 4, 8, 9, 11, 12, 13, 14 ≈ $9.00)

Shadow stick

The shadow stick is a pole with a stand that is used to determine the angle of incidence of the sun. It can also be used to determine actual solar time or to study seasonal shifts in the sun's declination.

Older children can use the shadow stick to determine the height of a tall building or tree by the application of similar triangles (fig. 14.17).

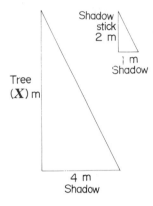

Fig. 14.17

Temperature

Blank thermometer

Children should experiment with non-standard units for each specific concept before standard units are introduced. This thermometer (fig. 14.18) allows children to experiment with gradations of their own choice before they are introduced to standard scales. (14 ≈ $0.50)

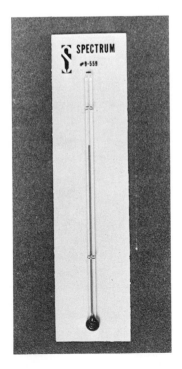

Fig. 14.18

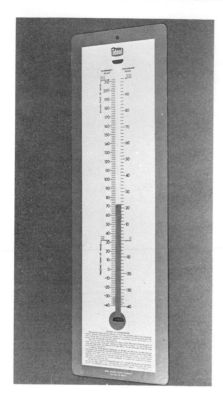

Student-lab thermometers

Children should learn about measuring temperatures by reading thermometers both in the home and in the school. They should be encouraged to make graphs of daily temperatures using both Fahrenheit and Celsius scales. The recorded temperature changes could guide the children to activities with signed numbers.

Thermometers are available with both Fahrenheit and Celsius scales (fig. 14.19). Children can thus observe temperature equivalences directly. (1, 2, 4, 6, 7, 8, 13, 14 ≈ $3.00)

Fig. 14.19

Area

Squares 10 cm × 10 cm

The set, which includes twenty square plastic tablets each 10 cm × 10 cm—or one square decimeter—is an excellent aid in developing the concept of area (fig. 14.20). Five sets would cover a square meter. (4 ≈ $4.00)

Fig. 14.20

Transparent centimeter grids

When used on an overhead projector, this transparent acetate grid (fig. 14.21) is ideal for showing children different areas. Each square unit is one square centimeter. A problem-solving activity could be to lay out simple closed curves on the overhead and have children determine the partial areas within the curve and combine them to get the total area. (1 ≈ $0.50)

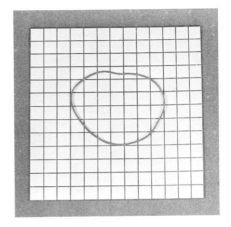

Fig. 14.21

One square yard

The concept of a square yard can be introduced with these nine square area cards (fig. 14.22), each of which covers one square foot. The cards can also be turned over and used to show that 144 square inches make up a square foot.

The activity leads to solving such practical problems as, How many square yards will carpet a room 9 feet by 10 feet? Another activity might be to determine how many square-inch tiles would be needed to tile a square yard. (7 ≈ $2.00)

Fig. 14.22

Perimeter/area board

The area board (fig. 14.23), a useful device to help children learn area and perimeter, is made of cork and has twenty large pins for marking off different areas. The cork is marked off in square inches. Children could be

Fig. 14.23

guided to discover that polygons of equal perimeters do not necessarily have equal areas and conversely that polygons with equal areas do not necessarily have equal perimeters. (7 ≈ $4.00)

Metric geoboard

This hard plastic geoboard has a grid 20 cm × 20 cm, with pins set at 2-cm intervals. One side has the pins set in a square pattern; the other side has a grid of equilateral triangles with 2-cm sides and pins set at each intersection. (See fig. 14.24.)

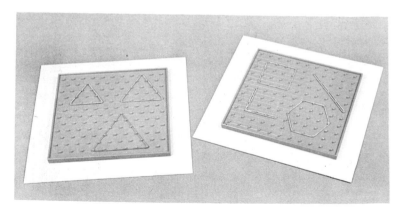

Fig. 14.24

The aid is excellent for introducing area and perimeter and helping to develop ideas of point, line segment, polygons, angles, and rays. (4 ≈ $2.50)

Fraction disks

The set of fraction disks in figure 14.25 is an excellent device for introducing partial areas. The set has color-coded disks made of six-inch cardboard pieces and includes halves, thirds, fourths, fifths, sixths, and eighths. Two plastic trays hold the fractional pieces. (10 ≈ $3.25)

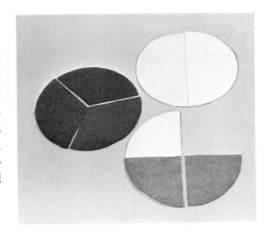

Fig. 14.25

Volume

Cubic meter

An outline of a cubic meter is made from twelve wooden dowels attached at the vertices as shown in figure 14.26. It is easily assembled and measurable by elementary school children and vividly demonstrates the magnitude of volume. (4, 6, 8, 14 ≈ $9.00)

Fig. 14.26

Centimeter cubes

Centimeter cubes (fig. 14.27) are an excellent illustration for the concepts of length, area, and volume. The cubes can also test a child's ability to conserve these concepts.

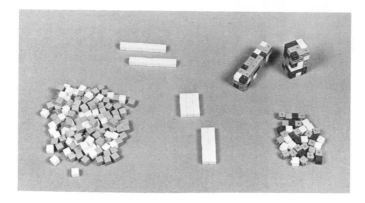

Fig. 14.27

Many different kinds of sets of centimeter cubes are on the market. Some sets contain pieces of different colors; others contain pieces all of one color. Some cubes have locks for making interlocking chains. (1, 2, 4, 5, 6, 8, 11, 14 ≈ $5.00 per 100 cubes)

Dissectible cubic foot

This device demonstrates concepts involving surface area and volume. The dissectible cubic foot (fig. 14.28) is made of cardboard and has many different subdivisions ranging in volume from 432 cubic inches to 1 cubic inch. All parts are scored with inch gradations on every side. (7 ≈ $5.00)

Fig. 14.28

Metric volume set

This volume set has a 10-ml, 100-ml, 250-ml, 500-ml, and 1-liter container (fig. 14.29). The containers are made of plastic, have pouring spouts, and are shatterproof. This set will permit a variety of pouring and measuring activities that will help children understand volumes and capacities. (1, 2, 4, 5, 6, 7, 8, 13, 14 ≈ $7.00)

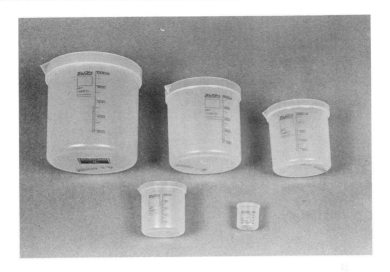

Fig. 14.29

5-ml spoons

These 5-ml spoons are made of plastic, are easy to level when measuring, and can provide the child with a very small measuring instrument. The spoon can be used to fill small containers with quantities like sand (fig. 14.30) and to find the capacity of a larger container. (4 ≈ $3.00 per 100)

Fig. 14.30

Graduated beakers

The plastic beakers in figure 14.31 have pouring spouts and include a 250-ml, a 500-ml, and a 1000-ml (one liter) beaker. Gradations on each beaker vary in size with respect to the quantity held. (1, 2, 4, 6, 8, 9, 13, 14 ≈ $11.00)

Fig. 14.31

Analysis of a liter

This set of soft plastic bottles (fig. 14.32) provides for a wide variety of experimental learning for elementary school children. It consists of ten 0.1-liter bottles, four 0.25-liter bottles, two 0.5-liter bottles, and one 1.0-liter bottle. (2, 4, 6, 8, 14 ≈ $10.00)

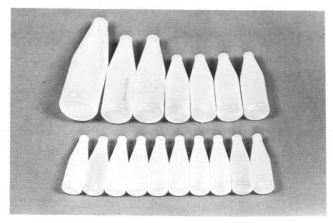

Fig. 14.32

Displacement pail

The displacement pail (fig. 14.33) can be used to demonstrate Archimedes' principle of volume displacement. Children can fill the pail with water to the bottom edge of the spout and place a graduated beaker under the spout to measure the water displacement of objects that are immersed in the pail. (2, 4 ≈ $4.00)

Fig. 14.33

Liter volume set

The set in figure 14.34 contains a cube, two cylinders of different diameters and heights, and a rectangular parallelepiped, all of which have a capacity of one liter. The containers are made of clear plastic with gradations on the outside. (1, 2, 4, 5, 8, 9, 14 ≈$8.00)

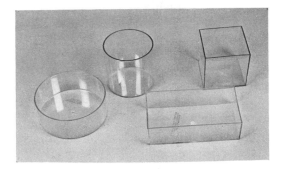

Fig. 14.34

Liquid measure

The set in figure 14.35 includes containers holding one gill, one pint, one quart, and one gallon. The plastic containers can be used for measuring solids (sand, flour) or liquids. Each container has a cover and a mark on the rim for accurate measuring. (7, 9 ≈ $8.00)

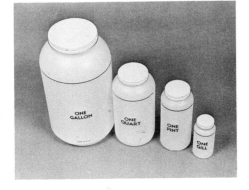

Fig. 14.35

Equal-volume geometric solids set

The plastic solids (fig. 14.36) are excellent for demonstrating how various-sized objects can displace the same amount of liquid. The set

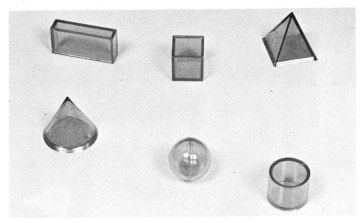

Fig. 14.36

contains six basic solids, each having a volume of one cubic inch: cube, pyramid, sphere, cone, parallelepiped, and cylinder. ($5 \approx \$3.00$)

Mass and Weight

The word *weight* as used here will refer both to kilograms, the measurement unit of mass, and to newtons, the measurement unit of force. The names of measurement units will be those in common usage for elementary school children.

One-piece balance

This balance has a calibrated beam, a fulcrum with a built-in level, and an adjustable weight on the counterbalancing end of the beam. In addition to introducing children to the notion of how a lever works, the balance presents them with a working model of an indirect way to measure weight (fig. 14.37).

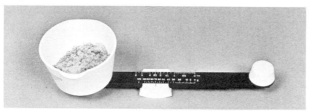

Fig. 14.37

This balance, which is suitable for weighing solids as well as liquids, has no loose parts and is light enough to be carried by young children. It is made of plastic and can be cleaned in water. (1, 2, 4, 14 $\approx$ $13.00)

Equal-pan balance

Children place the object they want to weigh in one pan of the balance and then proceed to find the equivalent mass for the other pan. The one shown in figure 14.38 is made of cast iron and has light steel pans in which

Fig. 14.38

to place the masses. Children should select several objects in the classroom and determine the weight of each. The balance can also be used for
nonstandard measures by letting the child select his own standard. (2, 4,
6, 8, 9, 11, 12, 14 ≈ $16.00)

Compression scales

The scale in figure 14.39 can be used with or without the plastic scoop
and is almost completely made of plastic, with the exception of the plat-
form and the internal springs. The calibrations range from 0 kg to 1 kg in
10-g divisions on the scale shown. For greater weights, others are available
in calibrations of 10 kg $\times$ 50 g or 5 kg $\times$ 20 g. (1, 2, 4, 6, 8, 9, 12,
14 ≈ $12.00)

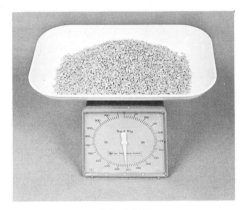

Fig. 14.39

Spring balances

An inexpensive form of a balance or
weight scale is the spring balance. This balance has a ring on the top so that it can
hang from a stationary beam. A hook on the
bottom is provided so that the object to be
weighed can be easily attached. The weight
range for available balances varies. Most
common are 3 kg with 50-g intervals (fig.
14.40), 6 kg with 100-g intervals, and 10 kg
with 250-g intervals. (2, 4, 6, 8, 12, 13,
14 ≈ $2.00)

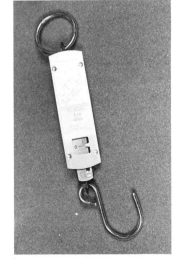

Fig. 14.40

Metric Place-Value Aids

Metric place-value chart

The place-value chart (fig. 14.41) will help teachers demonstrate the relationship between decimal expansion and the prefixes of the metric system. The chart has removable hangers for each position; it can be used at an elementary level and will permit conversion on both sides of the decimal. (7, 8 ≈ $4.00)

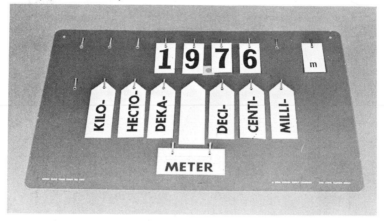

Fig. 14.41

1 kilo (base word)	1 thousand
1 hecto (base word)	1 hundred
1 deka (base word)	1 ten
1 (base word)	1 one
1 deci (base word)	1 tenth
1 centi (base word)	1 hundredth
1 milli (base word)	1 thousandth

base words: *meter, gram, liter*

Centimeter base-ten set

The centimeter base-ten set of blocks (fig. 14.42) consists of decimeter flats, longs, and cubes. The set has one cubic-decimeter block, ten flat blocks (10 cm × 10 cm × 1 cm), ten long blocks (10 cm × 1 cm × 1 cm), and one hundred cubes (1 cm on a side).

This versatile manipulative can be used to demonstrate volume concepts, base-ten numeration concepts, and metric-prefix concepts. Children should be encouraged to build different volumes with these blocks, determining the volumes built each time. Some sets of blocks are scored only on the top side; others are scored on all sides. (1, 2, 4, 6, 7 ≈ $30.00)

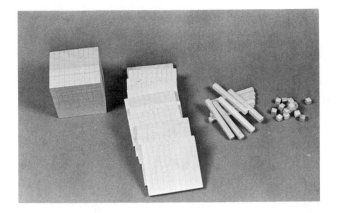

Fig. 14.42

DISTRIBUTORS

1. Creative Publications
   P.O. Box 10328
   Palo Alto, CA 94303

2. Creative Teaching Associates
   P.O. Box 7714
   Fresno, CA 93727

3. Cuisenaire Co. of America, Inc.
   12 Church Street
   New Rochelle, NY 10805

4. Dick Blick
   P.O. Box 1267
   Galesburg, IL 61401

5. Edmund Scientific Co.
   604 Edscorp Bldg.
   Barrington, NY 08007

6. Educational Teaching Aids
   159 W. Kinzie Street
   Chicago, IL 60610

7. Ideal School Supply Co.
   11000 S. Lavergne Avenue
   Oak Lawn, IL 60453

8. Math-Master
   P.O. Box 1911
   Big Spring, TX 79720

9. Math Shop Inc.
   5 Bridge Street
   Watertown, MA 02172

10. Milton Bradley Co.
    School Department
    P.O. Box 1581
    Springfield, MA 01101

11. Mind/Matter Corp.
    P.O. Box 345
    Danbury, CT 06810

12. Selective Education Equipment, Inc.
    3 Bridge Street
    Newton, MA 02195

13. Sigma Scientific, Inc.
    P.O. Box 1302
    Gainesville, FL 32601

14. Spectrum
    8 Denison Street
    Markham, Ontario L3R 2P2

15. The Math Group Inc.
    5625 Girard Avenue S.
    Minneapolis, MN 55419

16. The Instructor Publications, Inc.
    P.O. Box 6108
    Duluth, MN 55806

15

A Metric Bibliography

Gary G. Bitter
Charles P. Geer

The listing in this bibliography represents a sample of the software materials that are available for the teaching of measurement. It includes books and workbooks for both metrication and metric measurement. In addition to listing the publisher and the date of publication, each selection is annotated with a brief description of content, the number of pages, and an estimate of the level of the target audience according to the following code:

$$P = \text{Primary (grades K–3)}$$
$$E = \text{Elementary (grades 4–6)}$$
$$J = \text{Junior high school (grades 7–9)}$$
$$A = \text{Adult (grade 10–adult)}$$
$$T = \text{Teacher reference}$$

To make good use of this bibliography, teachers should peruse the selections to find sources of information that will provide for personal improvement and understanding or suitable measurement applications for their classrooms. We hope teachers will be able to use this listing with minimal effort and will find rich sources of measurement information to enhance the curriculum.

American National Metric Council. *Metric Conversion Paper.* 5 vols. Washington, D.C.: The Council, 1974. (A), 24 pp. each booklet.

> Set of booklets that discuss topics relating to the U.S. conversion to the metric system.

Armbruster, Frank O. *Think Metric.* San Francisco: Troubador Press, 1974. (E), 37 pp.

> Book designed to introduce students and teachers to the metric system through cartoons, designs, puzzles, and "put together" activities.

Barbrow, Louis E. *What about Metric?* Washington, D.C.: Government Printing Office, 1973. (A), 17 pp.

> Book explaining the metric system through pictures and diagrams.

Barnett, Carne. *Metric Ease.* Palo Alto, Calif.: Creative Publications, 1975. (E), 79 pp.

> Collection of activities designed to teach the facts of the metric system while giving students an opportunity to think metric.

Bitter, Gary G., and Thomas Metos. *Exploring with Metrics.* New York: Julian Messner, 1975. (P), 64 pp.

> Book introducing metric measurement to children with activities, related photos, and cartoons.

Bitter, Gary G., Jerald Mikesell, and Kathy Maurdeff. *Activities Handbook for Teaching the Metric System.* Boston: Allyn & Bacon, 1976. (T), 400 pp.

> Book for use by teachers as a reference for the metric system; includes history, activities, tests, and teaching suggestions.

————. *Discovering Metric Measure.* New York: McGraw-Hill Book Co., 1975. (E), 116 pp.

> Workbook for classroom use on the metric system; includes basic information, activities, and games related to the metric system. All related materials included in the book. Teacher's manual available.

————. *Investigating Metric Measure.* New York: McGraw-Hill Book Co., 1975. (J), 120 pp.

> Workbook that presents a discovery/involvement approach to the metric system using games, puzzles, and related activities; teacher's manual available.

Brandon, Julian R. *A Collection of Materials and Ideas on Metric Education.* Lansing, Mich.: Science and Mathematics Teaching Center, 1974. (T), 52 pp.

> Collection of articles on teaching measurement; includes an extensive bibliography.

Branley, Franklin. *Measure with Metric.* New York: Thomas Y. Crowell Co., 1975. (P), 33 pp.

> Book providing measurement activities for children.

Buckeye, Donald A. *Introducing the Metric System with Activities.* Troy, Mich.: Midwest Publications, 1972. (E), 26 pp.

> Collection of fifty activities written in the form of task cards for teaching the metric system.

————. *I'm O.K.—You're O.K. Let's Go Metric.* Troy, Mich.: Midwest Publications, 1973. (J), 62 pp.

Book of measurement activities involving direct measurement experiences using easily obtainable materials.

Buffington, Audrey V. *Meters, Liters, and Grams.* New York: Random House, 1974. (P, E), 64 pp. each book.

Series of six workbooks for introducing and teaching the metric system to elementary school students.

California State Department of Education. *In-Service Guide for Teaching Measurement (K–8): An Introduction to SI Metrics.* Sacramento, Calif.: The Department, 1975. (T), 48 pp.

Guide presents a model for measurement that takes into account the measurement concepts and logical development of the child.

Carr, Edwin. *Catalogue of Metric Instructional Materials.* Palo Alto, Calif.: American Institute for Research, 1974. (T), 55 pp.

Catalog that provides a list of suppliers of metric materials; lists measurement devices and provides list of suppliers, description, and cost of all products.

Cech, Joseph, and Carl Seltzer. *Working with Color Rods in Metric Measurement.* 3 vols, Skokie, Ill.: National Textbook Co., 1973. (E), 24 pp. each book.

Series of measurement books in spirit-master form; presents metric activities using colored rods.

Clack, Alice, and Carol Leitch. *Amusements in Developing Metric Skills.* Troy, Mich.: Midwest Publicaitons, 1973. (E), 47 pp.

Collection of games, puzzles, codes, and designs to develop and reinforce metric skills.

Cortese, Carol, and Carleton Herer. *Metric Measurement.* New York: American Books Co., 1974. (E), 60 pp.

Student workbook providing measurement activities; also includes a variety of activities on the metric system.

Deming, Richard. *Metric Power.* Nashville, Tenn.: Thomas Nelson, 1974. (A), 144 pp.

Book written in novel format explaining the history and development of measurement.

DeSimone, Daniel V. *A History of the Metric System Controversy in the United States.* Washington, D.C.: Government Printing Office, 1971. (A), 306 pp.

Book presenting a detailed historical account of the metric system controversy in the United States.

————. *A Metric America: A Decision Whose Time Has Come.* Washington, D.C.: Government Printing Office, 1971. (A), 170 pp.

A report to Congress prepared by the United States Department of Commerce; includes the history of the system, benefits and costs, and arguments for going metric. A major U.S. reference source on the metric system.

Donovan, Frank. *Let's Go Metric*. New York: Weybright & Talley, 1974. (A), 154 pp.

Brief introduction to the metric system, including its origin, functions, and uses.

Esler, William, and Michael Hynes. *An Individualized Laboratory Program for the Metric System*. Oviedo, Fla.: Kent Educational Services, 1973. (E), 138 pp.

Seven modules for teaching the metric system. Each module includes an activity book and laboratory manual.

Fillinger, Louis. *Know the Essentials of Metric Measurement*. 2 vols. Wilkinsburg, Pa.: Hayes School Publishing Co., 1974. (E), 18 pp. each book.

Series of books containing spirit masters on the mathematics of the metric system.

Geer, Charles, and John Geer. *MathMETRICS*. Burlingame, Calif.: Modern Math Materials, 1975. (E, J), 48 pp.

Series of activities to introduce the metric system and provide measurement activities for students; contains codes, games, and puzzles on the metric system.

General Motors Corporation (GMC). *The Swing to Metric*. Detroit: GMC, 1974. (A), 14 pp.

Free pamphlet dealing with the background, growth, principles, and impact of the metric system on General Motors and the automobile industry in general.

Gilbert, Thomas F., and Marilyn B. Gilbert. *Thinking Metric*. New York: John Wiley & Sons, 1973. (A), 140 pp.

Book designed as a self-instructional guide to the metric system; includes self-tests.

Glaser, Anton. *Neater by the Meter*. Southampton, Pa.: The Author, 1974. (A), 114 pp.

Informative book written for the teacher or layperson trying to understand the metric system; contains many examples of metric measurements and explains the application of important metric units.

Gould, Carole. *Color Me Metric*. San Jose, Calif.: A. R. Davis Co., 1973. (P), 15 pp.

Workbook that presents pictures and activities requiring students to use the metric system.

Great Ideas. *Metric Arithmapuzzles*. Commack, N.Y.: Great Ideas, 1973. (E), 22 pp.

Series of metric puzzles in crossnumber form requiring the use of various mathematical operations in the metric system.

Harder, Dale. *Metric Plus*. Castro Valley, Calif.: Education Plus, 1973. (E, J), 96 pp.

Book of metric activities and enrichment topics on measurement and the metric system; a metric kit designed for use with this book is available.

Hartsuch, Paul J. *Think Metric Now.* Chicago: Follet Publishing Co., 1974. (E), 120 pp.

    Book that presents activities and problems on the metric system; includes work on thinking metric as well as on conversion.

Haugaard, Jim, and David Horlock. *Contemporary Metrics.* Englewood Cliffs, N.J.: Prentice-Hall Learning Systems, 1973. (P), 30 pp.

    Book with activities on measuring, computing, and converting in the metric system.

————. *Fun and Games with Metrics.* Englewood Cliffs, N.J.: Prentice-Hall Learning Systems, 1974. (E, A), 94 pp.

    Book containing different activities and games with practice on the decimal and the metric system.

Henderson, George L., and Lowell D. Glunn. *Let's Play Games in Metrics.* Skokie, Ill.: National Textbook Co., 1974. (E), 223 pp.

    Over 170 objective-associated games and activities provided for various metric concepts; an introductory section presents basic information on the metric system.

Henry, Boyd. *Teaching the Metric System the Direct All Metric International Way.* Chicago: Weber Costello, 1974. (E, J), 48 pp.

    Activity book designed to tell the what, where, why, when, and how of the metric system through measurement experiences.

Hestwood, Diana, Anne Bartel, Ed Harter, Royce Helmbrecht, and Earl Orf. *Metric Measurement: Activities in Linear Measurement.* Minneapolis: Math Group, 1974. (J), 36 pp.

    Series of activities and puzzles for learning and reinforcing metric concepts of linear measurement.

Hiatt, Marty, and Linda Harvey. *The Metric Mice Measure.* 4 vols. Carson, Calif.: Educational Insights, 1974. (P), 28 pp. each book.

    Activity books to introduce concepts of measurement and the metric system using the laboratory and activity-card approach.

Helgren, Fred J. *Metric Supplement to Science and Mathematics.* Waukegan, Ill.: Metric Association, 1973. (J, A), 28 pp.

    Workbook of activities requiring laboratory work on the metric system and emphasizing the skills of estimating, measuring, converting, and problem solving.

Hirsch, S. Carl. *Meter Means Measure.* New York: Viking Press, 1973. (A), 126 pp.

    Resource book approaching the metric system from a historical standpoint and tracing the development of this system from its earliest beginnings to the present time.

Holmberg, Verda. *The Metric System of Measurement.* Hayward, Calif.: Activity Resources Co., 1973. (J), 98 pp.

    Activity workbook designed to provide enrichment materials for teaching the metric system.

Holt, Susan. *The United States and the Metric System.* Minneapolis: Federal Reserve Bank of Minneapolis, 1973. (A), 35 pp.

Free pamphlet on the metric system, the worldwide standardization of measurement, and its implications in world trade.

Higgins, Jon L., ed. *A Metric Handbook for Teachers.* Reston, Va.: National Council of Teachers of Mathematics, 1974. (A, T), 137 pp.

A collection of seventeen articles on introducing, teaching, and explaining the metric system; includes guidelines for the measurement process.

Hopkins, Robert A. *The International (SI) Metric System and How It Works.* Tarzana, Calif.: Polymetric Services, 1975. (A), 288 pp.

Provides a complete, detailed study of the metric system; includes sections on implications for education and educational aids.

Instructor Publications. *Practice in the Metric System.* Dansville, N.Y.: Instructor Publications, 1973. (E), 20 pp.

Twenty self-directing spirit masters introducing metric vocabulary and measurements.

Izzi, John. *Metrication American Style.* Bloomington, Ind.: Phi Delta Kappa, 1974. (T), 50 pp.

General information book on measurement that presents topics, recommended sources, and projects on the metric system.

J. J. Keller and Associates. *The Metric System Guide Library.* 5 vols. Neenah, Wis.: J. J. Keller & Associates, 1974. (A), 400 pp.

Series dealing with important metric topics that include definitions, regulatory controls, reference sources, metrication in the United States, and a metric units edition; books are updated by monthly mailings.

Leffin, Walter W. *Going Metric: Guidelines for the Mathematics Teacher, Grades K–8.* Reston, Va.: National Council of Teachers of Mathematics, 1975. (A, T), 48 pp.

Gives a history of the metric system; discusses the major SI units, with tips for teaching them; and suggests classroom activities, with recommended materials and instructions for student-made learning aids.

Long, Betty, and Carol Witte. *Fun with Metric Measurement.* Manhattan Beach, Calif.: Teachers, 1973. (E), 84 pp.

Activity book of games, codes, and puzzles to reinforce metric concepts.

May, Lola, and Donna Jacobs. *Metric Measurement.* 4 vols. New York: Harcourt Brace Jovanovich, 1974. (P, E, J), 32 pp. each booklet.

Series of workbooks that emphasize the metric system and the mathematical skills involved, particularly measurement and estimation.

McDonald's Corporation. *Fun Course in McMetrics.* Chicago: The Corporation, 1974. (P, E, J), 8 pp.

Free booklet presenting a metric unit on length and puzzle activities on this unit.

Michigan Council of Teachers of Mathematics. *Metric Measurement Activity Cards.* Birmingham, Mich.: The Council, 1974. (P, E), 72 pp.

Book set up in activity-card format containing a large number of metric activities; pages designed to be made into activity cards.

Miller, Mary R., and Toni C. Richardson. *Making Metric Maneuvers.* Hayward, Calif.: Activity Resources Co., 1974. (P, E), 106 pp.

Resource guide emphasizing measurement activities; pages designed to be used as activity cards.

National Council of Teachers of Mathematics. *The Metric System of Weights and Measures.* Twentieth Yearbook. New York: Bureau of Publications, Teachers College, Columbia University, 1948. (T), 303 pp.

Book of essays discussing uses of the metric system in education, science, athletics, and medicine in the late 1940s.

Orange County (California) Department of Education. *A Metric Workbook for Teachers of Consumer and Homemaking Education.* Anaheim, Calif.: The Department, 1973. (A, T), 62 pp.

Book for homemaking and consumer-education classes on teaching the metric system; includes suggestions for teaching and guidelines for setting up a metric workshop.

Orf, Earl, and Diana Hestwood. *Metric Measurement Activities in Capacity, Mass, and Temperature.* Minneapolis: Math Group, 1974. (J), 36 pp.

Activity book containing a series of activities on metric measurement using codes, shade-in puzzles, and graphs.

Page, Chester, and Paul Vigoureaux, eds. *The International System of Units.* Washington, D.C.: Government Printing Office, 1972. (A, T), 51 pp.

Reference booklet that presents the International System of measurement; gives definitions, multiples and submultiples, and symbols for metric units. Includes metric appendixes.

Parks, E. B. *Measuring à la Metric.* San Jose, Calif.: Parks & Math Co., 1973. (J), 60 pp.

Workbook on metric units emphasizing conversion to other metric units and the mathematics of the system.

Price, Shirley, and Merle Price. *Projects in Metric Measurement.* Monterey Park, Calif.: Creative Teaching Press, 1974. (J), 48 pp.

Book of self-directed activity cards for working with metric measurement.

Rabenau, Diane F. *The Metric System.* 3 vols. St. Louis: Milliken Publishing Co., 1974. (P, E), 28 pp. each booklet.

Series of spirit-master books developed for teaching the metric system; includes primary premetric skills through upper elementary work.

Ring, Arthur E. *Metrics.* Campbell, Calif.: Class Room Service Co., 1973. (P, E), 55 pp.

Resource book of teaching ideas and suggestions; includes information for interest centers and learning stations.

Robinson, John E. *How You Could Possibly Live with the Metric System.* Orange, Calif.: Human Design Associates, 1972. (A), 71 pp.

Book summarizing major forces involved in the U.S. conversion to metric measurement; includes influences in daily life, home, school, work, and travel.

Sears, Roebuck and Company. *An Educator's Guide to Teaching the Metric System*. Chicago: The Company, 1974. (T), 8 pp.

Free resource pamphlet for teachers explaining the metric system; suggests learning experiences and gives ideas for group projects on measurement.

Sherwood, J. C. *Discover Why Metrics*. South Bend, Ill.: Beloit Tool Co., 1972. (A), 65 pp.

Illustrated book that discusses reasons for using the metric system; includes a section on common conversions, style guides, and information on metric clothing sizes.

Smart, James R. *Metric Math: The Modernized Metric System*. Monterey, Calif.: Brooks/Cole Publishing Co., 1974. (A, T), 85 pp.

Resource book on the metric system; contains exercise sets and laboratory activities for a variety of units of measure.

Stern, Sidney. *Metric Games and Activities*. Paoli, Pa.: Instructo/McGraw-Hill, 1975. (E), 24 pp.

Book of spirit masters on games, puzzles, and activities related to the metric system.

Stover, Allan C. *You and the Metric System*. New York: Dodd, Mead & Co., 1974. (A, T), 95 pp.

Book explaining metric units, how they are used, how the changeover will affect the public, and what other problems will be involved in the conversion process.

Trueblood, Cecil R. *Metric Measurement*. Dansville, N.Y.: Instructor Publications, 1973. (P, E), 48 pp.

Activity booklet oriented to Piagetian activities for developmental stages; discusses the teaching of measurement and provides activities for teaching measurement topics.

Vogeli, Bruce R. *Metric Skills*. 6 vols. Morristown, N.J.: General Learning Corp., Silver Burdett Co., 1974. (P, E, J), 28 pp. each booklet.

Series of spirit-master booklets on measurement concepts and activities.

Wallace, Jesse D. *Going Metric . . . Pal*. Chico, Calif.: Jesse D. Wallace Co., 1974. (A), 41 pp.

Programmed-instruction booklet on metric measure for adults; includes metric information necessary for the worker, homemaker, and athlete.

Wallach, Paul. *The Metric Reader—a Metric Workbook*. Belmont, Calif.: Fearon Publications, 1975. (J), 112 pp.

Workbook to help students become familiar with the metric system; designed for students with a low reading level.

16

Low-Stress Algorithms

Barton Hutchings

$\mathbb{R}$ecently, an alleged decline in the computational skills of elementary and high school students has received a great deal of attention. Regrettably, part of this coverage has been sensational, which makes general understanding of the problem more difficult. Modern mathematics programs have received much of the blame, and a return to traditional programs is often viewed as the obvious solution.

Sources of the Present Difficulty in Teaching Computations

It is possible, without reviewing the strengths and weaknesses of individual programs, to offer a simple description of the historic forces that are producing the present difficulties. A primary cause is the rate at which human knowledge is expanding. A majority of all the scholars and scientists who have ever lived are alive and producing now. Their work expands the mathematical disciplines on which science and technology are built, thereby increasing the number of "quantitative" professions and vocations and making existing professions and vocations more quantitative. In order to retain a large portion of their career options, children must now learn much more mathematics, since there is much more mathematics to be learned. Unfortunately, there is no extra time or energy for learning it. The great increase in mathematical concepts and generalizations in the curriculum is fundamentally in conflict with the large amount of time and energy required for the mastery of conventional computation algorithms. Moreover, increased conceptual requirements in no way reduce the requirements for computational skill, particularly in applied professions and vocations. Arguments that the new curriculum develops better understanding of the operations are beside the point. Understanding division is not the same as knowing how to divide quickly and accurately. Each kind of learning, conceptual and computational, demands more time and attention than is now usually available. The conflict is real and critical, but it is basically the product of historic forces, not the fault of particular textbooks, teachers, or programs.

The Problem, and a Possible, Partial Solution

Fortunately, critics of mathematics education, whether journalists or lay people, often fail to realize the full dimensions of the problem. Poor computational skill, serious as that can be, is a small liability for students compared to feelings of anxiety and failure about mathematics. Such attitudes are certain to become worse unless the conflicting demands for curricular attention are resolved.

Moving the curriculum back toward traditional topics is not a solution —such a change would soon prove untenable and would do great damage. Clearer language, more effective lessons, and better assignment of concept priorities will help, but even after such reform the problem will remain and continue to grow. Instead, what is needed are drastic changes in the structure and teaching of computational skills.

Certain techniques, which are called low-stress algorithms, have been developed that might be a part of that change. They appear to permit easy mastery after brief training, to provide greater computational power than conventional algorithms, to operate with much less stress on the user than conventional algorithms, and to enjoy certain other advantages. After the procedures of low-stress algorithms have been demonstrated, their properties, history, lesson structures, theories, and applications will be discussed. The role of these procedures in the skills component of the mathematics curriculum may eventually be major and fundamental. It appears that immediate applications of the procedures should be directed toward students having severe remedial needs in the upper elementary and junior high school, as well as high school and adults.

Notation for Low-Stress Addition

$$\begin{array}{r} 8 \\ +\,7 \\ \hline 1\ 5 \end{array}$$

Conventional

$$\begin{array}{r} 8 \\ 7 \\ \hline 1\ \ {}_5 \end{array}$$

Half-space

The low-stress addition algorithm uses a new nota-tion, called half-space notation, to record individual steps. Half-space notation uses numerals of no more than a half-space in height to record the sum of two digits. With half-space notation, the units portion of the sum of two digits is written at the lower right of the bottom digit, and the tens portion is written at the lower left of the bottom digit. The example at the side shows both conventional and half-space notation.

Similarly,

$$\begin{array}{r} 7 \\ +\,5 \\ \hline 1\ 2 \end{array} \qquad \begin{array}{r} 8 \\ +\,2 \\ \hline 1\ 0 \end{array} \qquad \begin{array}{r} 6 \\ +\,9 \\ \hline 1\ 5 \end{array}$$

would be written this way:

$$\begin{array}{r} 7 \\ 5_2 \\ {}_1 \end{array} \qquad \begin{array}{r} 8 \\ 2_0 \\ {}_1 \end{array} \qquad \begin{array}{r} 6 \\ 9_5 \\ {}_1 \end{array}$$

Both the line and the plus sign are omitted in the new basic-fact notation, so that a slower child is not tempted to repeat the line or operation sign when doing column addition.

Naturally, when the sum is less than 10, no tens portion is recorded at the lower left. So

$$\begin{array}{r} 6 \\ +\,1 \\ \hline 7 \end{array} \qquad \begin{array}{r} 4 \\ +\,1 \\ \hline 5 \end{array} \qquad \begin{array}{r} 6 \\ +\,3 \\ \hline 9 \end{array}$$

would be written like this:

$$\begin{array}{r} 6 \\ 1_7 \end{array} \qquad \begin{array}{r} 4 \\ 1_5 \end{array} \qquad \begin{array}{r} 6 \\ 3_9 \end{array}$$

Single-column addition

When half-space notation is applied to column addition, changes in the usual procedure are possible. As with all low-stress algorithms, a complete record of component operations is made, and different kinds of operations can be completed separately without step-by-step alternation. This means that every basic addition fact necessary to the algorithm is recorded. Rather than recalling another addition fact and regrouping again, the student can recall and record all the necessary addition facts in an uninter-rupted sequence *and then* perform all the necessary regroupings. These regrouping operations, which are a major portion of the mental work of conventional algorithms, are drastically simplified by the new procedures.

Thus large sums can be efficiently obtained through a series of basic addition facts.

Consider this addition exercise:

$$
\begin{array}{r}
6 \\
8 \\
9 \\
4 \\
+\ 9 \\
\hline
\end{array}
$$

Starting at the top, if we add the first two digits, 6 + 8, and record the sum in the new notation, we have

$$
\begin{array}{r}
6 \\
8 \\
9 \\
4 \\
+\ 9 \\
\hline
\end{array}
$$

It is after this first step that the low-stress procedure becomes very different from the usual procedure: the sum 14 is *not* used in the next step. Instead, only the 4, the ones portion of the sum, is used. So, the next step is 4 + 9 = 13, and the 13 is recorded in our new notation:

$$
\begin{array}{r}
6 \\
8 \\
9 \\
4 \\
+\ 9 \\
\hline
\end{array}
$$

The rest of the work is done the same way.

The complete sum of each two-digit addition is recorded in half-space notation, but *only* the ones portion of each sum is used in the next addition:

$$
\begin{array}{r}
6 \\
8 \\
9 \\
4 \\
+\ 9 \\
\hline
\end{array}
$$

Work for the column is now complete. The ones portion of the column sum is *always* the same as the ones portion of the *last* two-digit sum—in this example, 6:

The tens portion of the column sum is *always* the same as the number of tens recorded at the left of the column. These are simply counted. Here there are three; so 3 is the tens portion of the column sum:

Consider the following examples. (Note that they are reprinted with each new step to show each step clearly to beginners. In actual practice, of course, students would *not* recopy the example for each step.)

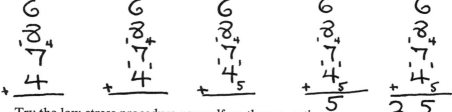

Try the low-stress procedure yourself on these practice exercises:

Your work should look like this:

Multicolumn addition

The low-stress procedures used for single-column addition can be applied without change to exercises involving more than one column. The advantage offered by low-stress procedures is increased in proportion to the length of the column. Columns should be spaced somewhat farther apart than with the usual procedure to allow space for the special notation. Carrying (or bridging, or regrouping, or whatever one might call it) can be done just as it is with the usual procedure. For a column in some multicolumn exercise, then, the last step—that is, counting the tens at the left of the columns—would be slightly changed. The counting itself is not changed in any way, but the answer, the total number of tens, is no longer written in the tens place of the first column's sum but instead at the top of the next column at the left. This is, in fact, exactly the same as the regrouping operation of the traditional procedures. The carried number is the tens total from the preceding column—in this example, 4. This value is used in the first two-digit addition of the column. This procedure could be extended to the addition of any multidigit whole number.

Consider the following example, which extends the previous one:

Work continues in this manner until the exercise is completed. Note, however, that the column sum for the *last* column in a multicolumn example is recorded in exactly the same way as the sum of a single-column exercise.

Here is another exercise to complete:

$$
\begin{array}{cccc}
6 & 5 & 3 & 5 \\
5 & 7 & 5 & 6 \\
8 & 6 & 2 & 7 \\
3 & 8 & 4 & 9 \\
6 & 9 & 3 & 8 \\
4 & 6 & 5 & 3 \\
+2 & 8 & 2 & 5 \\
\end{array}
$$

Your work should appear like this:

Locating specific errors

Specific errors can be located easily and without recitation. Consider these examples with a teacher's comments:

Chris

You always do 5 + 7 incorrectly. 5 + 7 = 12. Please write it 15 times and then talk to me. Also, you have carried incorrectly from the first column.

Kelly

You have correctly answered 7 + 5 twice, but you also missed it twice. You missed an easy fact, 1 + 3 = 4, which I'm sure you know. You also carried incorrectly from the first column. More careful work is needed.

A more complicated exercise

Extremely large problems can be solved with relatively little effort using low-stress procedures. This is true for all operations. Experience it now by working an example like the last one and then writing one of ten rows and six columns. Leave room for the new notation; don't be afraid to use space.

Solve the problem with the low-stress algorithm. Note that you can stop in the middle of a column and start again from that point.

Generating multiplication facts by low-stress addition

Generating multiplication facts is simple. The common factor of a desired table is written nine times in a column. For example, to generate the "seven times" table, a child would write this:

Note that the child is instructed *not* to write a line under the column and *not* to write a plus sign beside it. The omission of these symbols is considered an important part of directing the child's attention to the relevant partial sums.

After nine 7s are written in a column, the individual steps of low-stress addition are performed, but a complete column sum is *not* written:

The omission of the complete column sum is, like the omission of the bottom line and the plus sign, an important part of directing attention to relevant partial sums.

Now a child may read any multiplication fact by appropriately adjusting the column's length with a finger and regarding the exposed portion as an addition producing the desired fact. The product can be read as a sum expressed in low-stress notation.

$$3 \times 7 = 21 \qquad 6 \times 7 = 42 \qquad 9 \times 7 = 63$$

Generating division facts by low-stress addition

The procedure for generating division facts is very similar to the procedure for generating multiplication facts. The common factor is written nine times in a column and added as before; now, however, the tens portion of the dividend is counted at the left of the column downward, the units at the right of the column are matched to the number of tens, and then the repetitions of the column factor are counted back to the top of the column. Consider this example:

42 ÷ 7

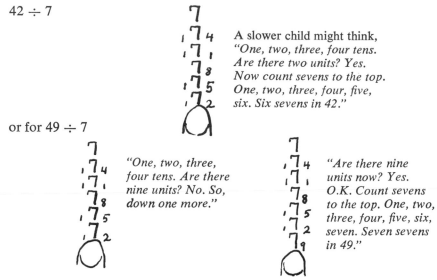

A slower child might think, *"One, two, three, four tens. Are there two units? Yes. Now count sevens to the top. One, two, three, four, five, six. Six sevens in 42."*

or for 49 ÷ 7

"One, two, three, four tens. Are there nine units? No. So, down one more."

"Are there nine units now? Yes. O.K. Count sevens to the top. One, two, three, four, five, six, seven. Seven sevens in 49."

Notation for Low-Stress Multiplication

Low-stress multiplication also has a special notational system. The notation uses a position drop of one row for the units-place numeral occurring in products of two single-digit factors. The technique is normally used with two partial products and not for simple multiplication facts. However, an easy way to learn the notation is to compare simple multiplication facts written in "drop" notation with the same facts written in conventional form. For example, the multiplication fact would be written in drop notation like this:

Similarly, the facts

would be written this way:

For consistency, the drop must be maintained even with the few multiplication facts that are not normally written with two-digit products. Thus the facts

are thought of as

and are written like this in drop notation:

$$\begin{array}{r} 2 \\ \times 3 \\ \hline {}_0 6 \end{array} \qquad \begin{array}{r} 4 \\ \times 1 \\ \hline {}_0 4 \end{array} \qquad \begin{array}{r} 3 \\ \times 3 \\ \hline {}_0 9 \end{array}$$

Multiplication with one multidigit and one single-digit factor

When drop notation is applied to multiplication exercises with one multidigit factor and one single-digit factor, all necessary multiplication facts are recorded and then all necessary addition operations are performed. The mental work involved in carrying in the conventional algorithm is virtually eliminated. Consider the following example and the accompanying steps.

$$\begin{array}{r} 3\ 6\ 8 \\ \times\ 7 \\ \hline \end{array}$$

Beginning in the usual direction, the right, we get these three steps:

$$\begin{array}{r} 3\ 6\ 8 \\ \times\ 7 \\ \hline \end{array} \qquad \begin{array}{r} 3\ 6\ 8 \\ \times\ 7 \\ \hline \ \ 5 \\ 6 \end{array} \qquad \begin{array}{r} 3\ 6\ 8 \\ \times\ 7 \\ \hline 4\ 5 \\ 2\ 6 \end{array} \qquad \begin{array}{r} 3\ 6\ 8 \\ \times\ 7 \\ \hline 2\ 4\ 5 \\ 1\ 2\ 6 \end{array}$$

Again, the entire binary product is recorded. The complete product is recorded with drop notation in a way that aligns the numerals of identical place value in each set of adjacent binary products.

Note that all the multiplication is completed before any of the addition is begun and that the multiplication is not interrupted by any other procedures. Note also that once the multiplication has been done, all the addition can be completed separately. (Low-stress addition could be used if needed. It is omitted here primarily to focus attention on low-stress multiplication.)

$$\begin{array}{r} 3\ 6\ 8 \\ \times\ 7 \\ \hline 2\ 4\ 5 \\ 1\ 2\ 6 \\ \hline 2\ 5\ 7\ 6 \end{array}$$

Use the following exercises for practice:

$$\begin{array}{r} 6\ 1\ 7\ 6\ 2 \\ \times\ 6 \\ \hline \end{array} \qquad \begin{array}{r} 4\ 0\ 6\ 2\ 7\ 1\ 5 \\ \times\ 9 \\ \hline \end{array}$$

Your solutions should appear like these:

$$\begin{array}{r} 6\ 1\ 7\ 6\ 2 \\ \times\ 6 \\ \hline 3\ 0\ 4\ 3\ 1 \\ 6\ 6\ 2\ 6\ 2 \\ \hline 3\ 7\ 0\ 5\ 7\ 2 \end{array} \qquad \begin{array}{r} 4\ 0\ 6\ 2\ 7\ 1\ 5 \\ \times\ 9 \\ \hline 3\ 0\ 5\ 1\ 6\ 0\ 4 \\ 6\ 0\ 4\ 8\ 3\ 9\ 5 \\ \hline 3\ 6\ 5\ 6\ 4\ 4\ 3\ 5 \end{array}$$

Right-left or left-right mode for recording multiplication

Experience indicates that it would be better to do all computations from left to right, for two reasons. First if we assume that the more operations we perform the more likely a mistake becomes, then in working from high place values to low, we shift the greater likelihood of error to the low place values. This should reduce the effects of error even when error frequency is constant. The error-probability structure of conventional algorithms tends to maximize the effects of error. Second, left-to-right operation can offer efficient and simple estimation procedures.

Experience also indicates that the left-to-right advantage in multiplication is such that only the left-to-right sequence should be used for problems having two multidigit factors. Experimental research to date, which will be described in later publications, has employed left-to-right low-stress approaches against right-to-left conventional approaches and found very significant advantages. However, the left-to-right option has not been tested independently of the low-stress advantage, partly because left-to-right operation is extremely awkward for the conventional algorithm.

Consider these same results obtained from the two modes:

RIGHT TO LEFT   *versus*   LEFT TO RIGHT

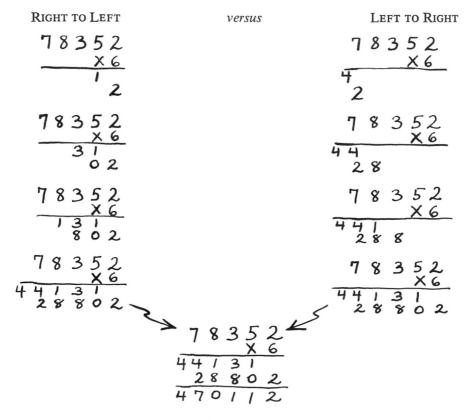

Using low-stress multiplication from left to right is simply a matter of starting at the left side. No other adjustment is required.

Multiplication with two multidigit factors

Proper place alignment in the left-to-right operation appears easy (possibly because the direction of hand and eye movement is the same as in reading and writing). Consider:

$$7\ 8\ 7\ 6\ 3$$
$$\times\ \ 7\ 6\ 8$$

If we begin with the leftmost numerals of both factors and multiply left to right, we have the following:

$$7\ 8\ 7\ 6\ 3$$
$$\times\ 7\ 6\ 8$$
$$4\ 5\ 4\ 4\ 2$$
$$9\ 6\ 9\ 2\ 1$$

Note that we recorded the first binary product in a position three places to the left of the larger factor. This was done in order to align the units place in the completed products with the units place in the factors. Such alignment is commonly practiced, although it is not required for either mathematical correctness or easy work. Its only advantage is a small claim to neatness. The rule is that the units places of the completed product and the factors will be aligned if we begin by recording the partial products in a place as many places to the left of the leftmost numeral of the larger factor as there are digits in the lesser factor. When using this rule, remember that all products are regarded as occupying at least two places.

Partial products of the second digit of the bottom factor are always begun directly beneath the leftmost digit of the lower row of the previous double row. Although it is not necessary, a small horizontal mark can be used to denote this position:

$$7\ 8\ 7\ 6\ 3$$
$$\times\ 7\ 6\ 8$$
$$4\ 5\ 4\ 4\ 2$$
$$9\ 6\ 9\ 2\ 1$$

The second set of binary partial products moves left to right in the same manner as the first set:

$$7\ 8\ 7\ 6\ 3$$
$$\times\ 7\ 6\ 8$$
$$4\ 5\ 4\ 4\ 2$$
$$9\ 6\ 9\ 2\ 1$$
$$4\ 4\ 4\ 3\ 1$$
$$2\ 8\ 2\ 6\ 8$$

The partial products of the third digit of the bottom factor begin in the same relative position and move left to right in the same way. If added in the usual way, the completed example would look like this:

$$7\ 8\ 7\ 6\ 3$$
$$\times\ 7\ 6\ 8$$
$$4\ 5\ 4\ 4\ 2$$
$$9\ 6\ 9\ 2\ 1$$
$$4\ 4\ 4\ 3\ 1$$
$$2\ 8\ 2\ 6\ 8$$
$$5\ 6\ 5\ 4\ 2$$
$$6\ 4\ 6\ 8\ 4$$
$$60{,}4\ 8\ 9{,}9\ 8\ 4$$

Note that the number of numerals processed in this procedure is exactly the same as the number processed in the conventional procedure. However, the low-stress procedure provides a complete record of those numerals and eliminates almost all their mental manipulation.

Consider the following example. It is entirely repeated with each row of work to illustrate the procedure for beginners.

```
  74013          74013          74013
   X43            X43            X43
 21001          21001          21001
 86042          86042          86042
                21000          21000
                12039          12039
                             3,182,559
```

Now attempt this one:

```
385926
X3749
```

If you solved the problem on lined paper and used the lines to keep the columns straight, your work should look like this:

```
          385926
          X3749
 0 2 1 2 0 1
 9 4 5 7 6 8
 2 5 3 6 1 4
   1 6 5 3 4 2
   1 3 2 3 0 2
     2 2 0 6 8 4
     2 7 4 8 1 5
       7 2 5 1 8 4
 1,4 4 6,8 3 6,5 7 4
```

Locating specific errors

Specific errors can be located easily and without recitation. Consider these examples and some hypothetical teacher's comments to the student:

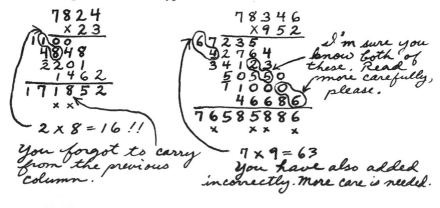

Note that a teacher could also classify types of error for diagnostic work. Every error is shown and is therefore accessible from the student's worksheet.

Notation for Low-Stress Subtraction

Low-stress subtraction has a special notation used to record the steps performed in solving a subtraction problem. The notation consists of two parts. The first is to record all upward regrouping of places by a half-space "1" placed at the upper left of numerals occupying such places. For example, **6 43** could be recorded in regrouped form as **5 '43** or **5 '3 '3** or **6 3 '3**.

The second is to write the regrouped minuend directly above its subtrahend, which helps some children in reading and organizing their work. The regrouped minuend is written "in the middle" rather than at the top, as shown in the following example:

$$6\ 2\ 8\ 3\ 4$$
$$\boxed{}$$
$$-\ 1\ 2\ 9\ 9\ 7$$

Subtraction without regrouping zeros

In subtraction, the entire minuend is regrouped and recorded *before* any subtraction occurs. For example, the exercise

$$8\ 4\ 7\ 2$$
$$-\ 6\ 6\ 7\ 3$$

is regrouped like this:

$$
\begin{array}{c}
8\ 4\ 7\ 2 \\
6\ '2 \\
\hline
6\ 6\ 7\ 3
\end{array}
\qquad
\begin{array}{c}
8\ 4\ 7\ 2 \\
3\ '6'2 \\
\hline
6\ 6\ 7\ 3
\end{array}
\qquad
\begin{array}{c}
8\ 4\ 7\ 2 \\
7\ '3'6'2 \\
\hline
6\ 6\ 7\ 3
\end{array}
$$

After this regrouping, all subtraction is completed in an uninterrupted sequence. Continuing with the previous example, we can begin on the right with $12 - 3$, and progress to a result. Thus

$$
\begin{array}{c}
8\ 4\ 7\ 2 \\
7\ '3\ '6\ '2 \\
\hline
6\ 6\ 7\ 3
\end{array}
\qquad \text{becomes} \qquad
\begin{array}{c}
8\ 4\ 7\ 2 \\
7\ '3\ '6\ '2 \\
6\ 6\ 7\ 3 \\
\hline
1\ 7\ 9\ 9
\end{array}
$$

Try the following subtractions, using the low-stress algorithm:

$$
\begin{array}{c}
4\ 2\ 6\ 3\ 5 \\
-\ 2\ 4\ 1\ 7\ 8
\end{array}
\qquad\qquad
\begin{array}{c}
6\ 3\ 5\ 4\ 7\ 2 \\
-\ 3\ 7\ 6\ 6\ 7\ 3
\end{array}
$$

The results should appear like these:

$$\begin{array}{r} 4\,2\,3\,6\,5 \\ 3\,'2\,2\,'5\,'5 \\ -\,2\,4\,1\,7\,8 \\ \hline 1\,8\,1\,8\,7 \end{array} \qquad \begin{array}{r} 6\,3\,5\,4\,7\,2 \\ 5\,'2\,'4\,'3\,'6\,'2 \\ -\,3\,7\,6\,6\,7\,3 \\ \hline 2\,5\,8\,7\,9\,9 \end{array}$$

Subtraction with regrouping zeros

The conceptual requirements for regrouping can, when necessary, be greatly reduced under the low-stress procedures. Moreover, since the low-stress algorithm is a decomposition algorithm, it is conceptually consistent with the conventional algorithm.

Consider the problem
$$\begin{array}{r} 7\,6\,3\,0\,0\,0\,0\,7\,2 \\ -\,2\,1\,4\,6\,0\,1\,3\,8\,5 \end{array}$$

The renamed minuend is begun exactly as before:
$$\begin{array}{r} 7\,6\,3\,0\,0\,0\,0\,7\,2 \\ 6\,'2 \\ -\,2\,1\,4\,6\,0\,1\,3\,8\,5 \end{array}$$

When regrouping becomes necessary, a child can simply ignore the zeros and proceed as before, moving on to the first number that allows borrowing, completing the portion of the renamed minuend that has no regrouped zeros:

$$\begin{array}{r} 7\,6\,3\,0\,0\,0\,0\,7\,2 \\ 2 \qquad\ 6\,'2 \\ -\,2\,1\,4\,6\,0\,1\,3\,8\,5 \end{array} \qquad \begin{array}{r} 7\,6\,3\,0\,0\,0\,0\,7\,2 \\ 5\,'2 \qquad\ 6\,'2 \\ -\,2\,1\,4\,6\,0\,1\,3\,8\,5 \end{array} \qquad \begin{array}{r} 7\,6\,3\,0\,0\,0\,0\,7\,2 \\ 7\,5\,'2 \qquad\ 6\,'2 \\ -\,2\,1\,4\,6\,0\,1\,3\,8\,5 \end{array}$$

Each ignored place, or "skipped zero," may then be replaced in the renamed minuend by 9:
$$\begin{array}{r} 7\,6\,3\,0\,0\,0\,0\,7\,2 \\ 7\,5\,'2\,9\,9\,9\,9\,6\,'2 \\ -\,2\,1\,4\,6\,0\,1\,3\,8\,5 \end{array}$$

This option substantially reduces the conceptual loading that makes subtraction a special obstacle to students having moderate to severe learning disabilities. Note that this mechanical treatment is presented primarily because it meets the urgent need of such remedial populations. Excellent development of meaning is possible because of the explicitness of every step; however, the author believes that performance should be emphasized for the student whose applications are likely to be practical.

Locating specific errors

Specific errors in low-stress subtraction can be located easily and without recitation. Consider these examples:

Please see me about these,
17 − 9 = 8 *and*
11 − 6 = 5

all differences of individual subtractions will have single-digit answers; if you get two digits, then you have regrouped when you did not need to.

Low-Stress Division

The low-stress division algorithm, like the conventional division algorithm, is a cycling of partial-quotient estimation, multiplication, and subtraction. Moreover, the components of the low-stress division algorithm are independent and may be used separately or in whatever combination desired.

The algorithms might employ low-stress multiplication and low-stress subtraction. The example

might be completed like this:

```
                    7
     256│2 0 4  1  1
   1 3 4    1 9 '3 '1
     4 5 2 1  7 9 2
   ┌1 7 9 2┐
   └       ┘
```

```
                      7
       256│2 0 4  1  1
     1 3 4    1 9 '3 '1
       4 5 2 1  7 9 2
     ┌1 7 9 2┐ 2 4 9
     └       ┘
```

```
                        7
         256│2 0 4  1  1
       1 3 4    1 9 '3 '1
         4 5 2 1  7 9 2
       ┌1 7 9 2┐ 2 4 9  1
       └       ┘
```

The same procedures are employed to obtain the quotient's second significant figure.

```
                          7 9      1 8 7
         256│2 0 4  1  1          2 5 6
       1 3 4    1 9 '3 '1
         4 5 2 1  7 9 2
       ┌1 7 9 2┐ 2 4 9  1
       1 4 5    2 4 8 '1
         8 5 4  2 3 0 4
       ┌2 3 0 4┐   1 8 7
       └       ┘
```

When a partial quotient is underestimated, the underestimate can be used in a manner similar to that of a better estimate, even in the conventional algorithm. Consider these examples:

```
                4 }           2 }          1 }
        8       4 } 8         6 } 8        7 } 8
     7│5 6    7│5 6        7│5 6        7│5 6
     5 6       2 8          4 2          4 9
              2 8          1 4            7
              2 8          1 4            7
              2 8
```

It is probably more efficient to proceed from such an underestimate than to discard the work and begin again. Moreover, the knowledge that any conservative estimate will be usable and will tend to simplify the next step appears to reduce anxiety. The technique is consistent with the general character of low-stress algorithms, and along with an inclination to make conservative estimates, it is suggested as a third component of low-stress division. An example in which the procedure is employed might look like this:

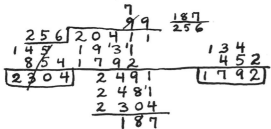

(Note that no regrouping is needed in the final subtraction.)

An overestimate can never be incorporated directly, of course, but we can enjoy the efficiency of coherently preserving the information it produces. The recommended procedure is to draw a thin line through both the incorrect quotient and the incorrect partial product, otherwise leaving them as written. The new estimate is written above the old estimate, and the new partial product is written to the *right* of the problem. This procedure is recommended because the information generated by the incorrect estimate might be needed later in the problem or in a subsequent problem; if available, it would not have to be regenerated. An exercise in which this occurred might look like this:

Multiplication by the units portion of the quotient was not needed, since relevant significant figures were obtained in the overestimate.

It is possible to locate the place of the quotient by a mechanical, non-conceptual procedure. The procedure requires simply that a finger be slid along the dividend until a numeral that exceeds the divisor is shown. The quotient always begins directly above the last digit of this numeral. Consider:

256⟌2̲0̲   256⟌2̲0̲0̲   256⟌2̲0̲4̲   256⟌2̲0̲4̲1̲ *

A numeral larger than the divisor appears at the fourth movement. The quotient begins directly above the last digit of this larger numeral.

Another example:

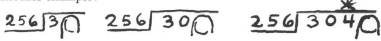

> A numeral larger than the divisor appears at the third movement. The
> quotient begins directly above the last digit of this larger numeral.

In each of the other algorithms, we noted that specific errors could be
located without recitation as each step is recorded. Obviously, this property
remains in low-stress division when all options are exercised.

Applications and the Development of Meaning

Although, as stated earlier, the low-stress procedures eventually may
have an important role in the revision of the standard skills curriculum,
their most urgent application is for that population of students having
extreme remedial needs. Someday these procedures might be standard
algorithms. These algorithms, together with hand-held electronic calculators, might form the skills core. Indeed, low-stress algorithms and hand-
held calculators complement each other and are a logical team. The
calculator offers speed with complex operations; the algorithms offer
independence from the machine, the power to check the machine easily
or even to exceed the machine's limitations, and a permanent, complete
record of work. Facility in both and the option to use either as needed
probably constitute the ideal skills package, necessitating only a fraction
of the time and stress required by conventional algorithms.

However, attractive as such prospects are, they have nothing to do with
immediate applications of the procedures. All of us should be aware of
the hazards of changing the curriculum too rapidly. Time must be allowed
for appropriate evolution. A possible intermediate step in such evolution
is the use of low-stress procedures as supplements, or sometimes as options,
to conventional algorithms. For each operation, there are sequences of
algorithmic modification allowing a progression in small steps from
expanded notation to low-stress notation to conventional notation. It may
be that such sequences could both assist in the development of certain
concepts and allow a relatively stress-free approach to conventional procedures. It appears that present applications should focus on children having
remedial needs but should not be introduced prior to grade 4. Children
in special-education classes are prime candidates for instruction with low-
stress algorithms. The failure of this population with conventional procedures is so broad that low-stress procedures may be their only real hope
for attaining competence. Other prime targets are high school remedial
classes and vocational training classes where students require a simple

numerical competence in order to be realistic candidates for certain kinds of employment. Any student in any class who is frustrated or failing with conventional procedures should be offered the low-stress techniques. However, priorities might be established, since the more students are impaired or the older they are, the less likely effective remediation through conventional methods becomes.

The development of the meaning of operations is beyond the scope of this essay, which describes only the mechanics. Whatever procedures a teacher uses to develop the meanings of the operations for conventional algorithms, expanded algorithms, the abacus, or whatever can also be used with low-stress algorithms. There is no shortage of such techniques.

Properties and Lesson Structures

Research on the low-stress algorithms indicates that they have three major effects on performance. First is a large reduction in the time required for mastery—that is, given some computational criterion in the moderate or difficult range, students using the low-stress algorithm will meet that criterion in much less time than students using the conventional algorithm. The second effect is a large increase in computational power. The third effect is a sharp reduction of the stress that occurs when challenging computations are performed. This effect is subjective but is easily experienced by doing a challenging example twice, once with a low-stress algorithm and once with a conventional algorithm. These effects appear to derive from two distinct cognitive characteristics, which, in turn, appear to derive from two mechanical characteristics. The low-stress algorithms have two basic mechanical characteristics: (1) a definable notation that allows concise records of all binary operations, and (2) the option for the separate completion of distinct sets of binary operations. For example, a multiplication exercise worked with the low-stress algorithms records all binary multiplications and all binary additions. Moreover, all binary multiplications could be completed and then all binary additions could be completed, rather than alternating between individual binary multiplications and individual binary additions as required in the regrouping process of the conventional algorithm. Similarly, a complete record can be made of fact recall in addition and subtraction, and afterwards regrouping operations of both can be completed separately from fact recall.

This complete record may produce the first distinct cognitive characteristic, which is a great reduction in the demands on memory or imaginary manipulation, compared to those made by conventional algorithms. The separation option may produce a second distinct cognitive characteristic, which is the reduction of dissonance and retroactive inhibition derived from the alternation of operations in conventional algorithms. Together, these characteristics seem to constitute virtual elimination of the mental work necessary with conventional algorithms.

Classroom experience has shown that initial instruction should focus on the algorithms' two mechanical distinctions. First, the notation should be demonstrated and practiced as it would be used with individual number facts. Then, after the various fact expressions are clear, the relationships of the notation to problem procedures can be demonstrated. Practice should initially focus on exercising the option to complete distinct sets of different kinds of operations separately. For example, addition practice should first involve long, single-column problems rather than shorter, multicolumn problems. Multiplication practice should first involve a factor of one digit and another factor of several digits rather than two factors each having two or three digits.

After the separation option is clear, various problem structures can be practiced. The great reduction in mental work makes exercises involving large multidigit numbers relatively easy, and assignments can progress quickly to such exercises so that students appreciate the dimension of their new computational power. There are two major advantages in using large problems: The one advantage is psychological. Children who can easily and correctly find results of seemingly complex exercises will have an effective success experience; they will be entertained by the challenge, and be much less intimidated by typical exercises on examinations or in textbooks. Another major advantage is practical power. The modest size of conventional computation exercises is due more to the limited power of conventional algorithms than to the dimension of computational demands that will really be made in the world, particularly in applied quantitative professions and vocations.

History and Theory

Research on low-stress algorithms began at the Syracuse University Arithmetic Center in 1967 and has been a major concern of the University of Maryland's Arithmetic Center since 1971. A great deal of independent work has also been done, and a large set of personal and pedagogical procedures has evolved.

The first algorithm to be developed, the beginning of the low-stress thrust, was the multiplication algorithm, which resulted from an unrelated need to employ lattice multiplication without the considerable hindrance of drawing a lattice. Inspiration for the addition algorithm came from two sources: watching an unknown carnival player do tricks with notationless modular addition at a state fair, and frequent unsuccessful attempts to balance an old-fashioned attendance ledger.

The subtraction algorithm began as an assignment to teach basic arithmetic to children grouped essentially on the basis of reading disability. With low-stress subtraction and multiplication available, low-stress division was an obvious inference. The procedures for generating facts were developed last as the distinct character of basic fact needs became apparent. All the algorithms were developed by the author, but a large team of colleagues and students are still involved in research on them and in perfecting instruction and training procedures.

Much research has been completed; reports and summaries will appear in various later publications. However, a vast amount of work remains to be done, both in the analysis of why the associated cognitive and affective events occur and in the more immediate areas of how well they work with certain types of learners. Collaboration is invited. Interested persons should write to Correspondence Coordinator, Low-Stress Research, Dr. James Potterfield, Education Chairman, Francis Marion College, Florence, South Carolina 29501.

Index